Neues verkehrswissenschaftliches Journal

Ausgabe 30

Track Data-Oriented Maintenance Intervention Limit Determination for Ballasted Light Rail Tracks through Multibody Simulations

Von der Fakultät Bau- und Umweltingenieurwissenschaften

der Universität Stuttgart zur Erlangung der Würde eines

Doktors der Ingenieurwissenschaften (Dr.-Ing.) genehmigte Dissertation

Vorgelegt von

David Camacho Alcocer

aus Naucalpan, Mexiko

Hauptberichter: Prof. Dr.-Ing. Ullrich Martin

Mitberichterin: Prof. Dr.-Ing. Corinna Salander

Tag der mündlichen Prüfung: 30. September 2019

Institut für Eisenbahn- und Verkehrswesen der Universität Stuttgart

2019

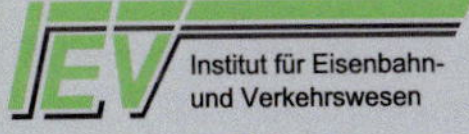

Herstellung und Verlag: BoD – Books on Demand, Norderstedt

Printed in Germany

ISBN 978-3-7534-6306-3

Preface

The quality of a railway track describes the correct position and the state of the track body. Both of these aspects change during track use and are particularly susceptible to change immediately after new construction or maintenance activities. In addition to ensuring safety during operation, maintaining a certain level of track quality also has a significant impact on the life cycle costs and availability of the railway infrastructure.

If the quality of the track decreases too much, a disproportionately high amount of maintenance will be required. However, if maintenance is carried out each time a slight drop in the track quality occurs, the number of maintenance interventions and the number of resulting restrictions to operations will increase disproportionately. Therefore, determining the adequate time for maintenance interventions and being able to reliably determine the quality of the track while maintaining normal rail operations are of central importance.

In his dissertation, Mr. Camacho develops a methodical approach to quasi-continuously determine the geometrical quality of the track. His approach takes into account the specific framework conditions for light rail systems so that normally operating vehicles can be used to record measurement data during their operation. Of particular interest in this work is the introduction of the standard BS EN 12299: 2009, a standard normally used for determining traveler comfort, as the basis for modeling the track quality. This new approach demonstrates the complexity of this research topic due to, among other reasons, the classification of different irregularities in the track quality and the filtering out of vehicle-related influences.

The main result of this research is a methodology for recording, evaluating, and interpreting measured values in order to determine the geometrical track quality in relation to vehicle responses with particular respect to minimizing the life cycle costs of light rail system infrastructure by implementing vehicle floor sensors in normal vehicles during normal operations, and by deriving the construction-related track quality based on comfort parameters.

Stuttgart, February 2020

Ullrich Martin

Dedication

To my family, especially my wife and daughters... for a brighter future…

Acknowledgment

First of all, I would like to offer my most grateful thank you to my wife and children. Without their support, sacrifice and effort not a single word in this document would have been possible. Thank you, my loved ones.

Without a doubt, my extended family upholds great merit in keeping the train running. I would specifically like to thank Maria Luigia Scuderi and Winfried Zimmermann, and my own family, my mother, Maria Eugenia Alcocer Maycotte, my father, Mario Camacho Cardona, and my sister Ruth.

Naturally, I would also like to thank Professor Ullrich Martin and Professor Corinna Salander for their great support during these past five and a half years, either through advice or resource provision. My thank you for knowledge and resource support could not be complete without thanking colleagues of both institutes, IEV and IMA. I thank all members for their comments and great discussions. However, there are specific people I would like to mention. From IEV a specially thank to Sebastian Rapp, who reviewed the document; providing important insight and comments. He also pushed me through until this document was submitted. From IMA, I would like to thank Timo Strobel in a very special manner. Timo in fact is greatly responsible for the development of the work through his active involvement, support and discussions.

This acknowledgement would not be complete without including some special people outside my family and academics. Hence, I would also like to thank all my friends and students who through the last five plus years have provided me with, in one way or another, support or advice, especially, Giovacchino Genesio, the Family Volz (esp. Michael and Beatriz), The family Rhein (esp. Dieter and Christine), Hanoch Serebrenik, Gabriela Vega and Maria Luisa Flores.

Last but not least, I would like to thank the Mexican government for the financial support which made it possible to dedicate all this time for the development of this research.

Eidesstattliche Erklärung

Hiermit erkläre ich, dass ich diese Arbeit selbständig verfasst und keine anderen als die von mir angegebenen Quellen und Hilfsmittel verwendet habe.

Stuttgart, den 30.09.2019   David Camacho Alcocer

Table of Contents

List of Figures

List of Tables

Kurzfassung

Stadtbahnen (Light rail trains, LRT) bilden einen wichtigen Teil des öffentlichen Personennahverkehrs. Aufgrund der hohen Lebenszykluskosten werden Stadtbahnsysteme jedoch nicht immer als geeignete Lösung betrachtet. Eine Möglichkeit die Lebenszykluskosten zu senken, besteht im Bereich des Instandhaltungsmanagements.

Die Instandhaltung von Stadtbahnstrecken basiert auf Erfahrungswerten der Anlagenverantwortlichen und fokussiert auf die Wartung und die fehlerbehebende Instandhaltung. Dabei wird der aktuelle Streckenzustand nicht immer ausreichend berücksichtigt, wodurch kaum Möglichkeiten zur Kostenoptimierung vorhanden sind. Insbesondere für Stadtbahnsysteme geltende Toleranz-und Grenzwerte für Instandhaltungsmaßnahmen sind bisher weder festgelegt noch ausreichend analysiert, weshalb Grenzen aus Richtlinien der Eisenbahn übernommen werden und dadurch angemessene Intervalle für Wartung- und Instandsetzungsmaßnahmen nicht erreicht werden.

Toleranz- und Grenzwerte für Stadtbahnsysteme sollten auf dem Fahrzeugverhalten basieren, welches wiederum von der aktuellen Gleislagequalität abhängig ist. Dies würde dauerhaft zu einer Gleislage führen, die entweder einen für den Fahrgast komfortablen oder einen für den Infrastrukturbetreiber wirtschaftlichen Betrieb erlaubt.

Der Schwerpunkt dieser Arbeit liegt auf der Bewertung der Gleisgeometrie von Stadtbahnstrecken, für welche, basierend auf der Fahrzeugreaktion, Toleranz- und Grenzwerte für Instandhaltungsmaßnahmen ermittelt werden. Da Daten für Streckenabschnitte in sehr schlechtem Zustand nicht vorhanden waren, werden durch eine sukzessive Erhöhung der Fehleramplituden innerhalb des Frequenzbereichs der gemessenen Gleisgeometrien, künstlich schlechte Gleislagen erzeugt. Die Gleismessdaten werden in eine Mehrkörpersimulationssoftware importiert und die Fahrzeugreaktion bei der Überfahrt von ausgewählten Streckenabschnitten für eine Fahrdauer von fünf Minuten berechnet. Anschließend wird anhand des Fahrkomforts, der Gleisbelastung sowie des Entgleisungskoeffizienten die Fahrzeugreaktion bewertet und auf die Qualität der Gleislage geschlossen.

Zur besseren Darstellung der von der Fahrzeugreaktion abgeleiteten Toleranz- und Grenzwerte für Stadtbahnsysteme wird ein Gleisgeometrieindex eingeführt. Der Gleisgeometrieindex ermöglicht die Darstellung der Verschlechterung der Gleislage und die

Einteilung dieser in unterschiedliche Qualitätsstufen. Die Qualitätsstufen helfen Anlagenverantwortlichen, den Zeitpunkt für Instandsetzungsmaßnahmen zielgerichtet festzulegen. Es werden zwei Stadtbahnstrecken detailliert mit der hier entwickelten Methode untersucht. Das Ergebnis deutet darauf hin, dass die in dieser Arbeit betrachteten Strecken zu früh instandgesetzt wurden.

Zukünftige Arbeiten sollen Gleislagefehler durch kontinuierlich gemessene Beschleunigungen am Fahrzeug und im Gleis detektieren. Zusätzlich können Beschleunigungsdaten dann auch für die Beurteilung längerer Streckenabschnitte mit dem hier entwickelten Ansatz genutzt werden.

Abstract

Light rail trains (LRT) form an important part of public transport. However, due to the high life-cycle costs, light rail systems are not always considered a suitable solution. One way to reduce life cycle costs is in the area of maintenance management.

The maintenance of light rail tracks is based on the experience of the infrastructure managers and focuses on preventive and corrective maintenance. The current track condition is not always sufficiently considered, which means there are hardly any possibilities for optimizing costs. In particular, tolerance and limit values for maintenance measures applicable to light rail systems have not yet been defined or adequately analyzed. Instead, limits are taken from the regular railway guidelines, which means that adequate intervals for maintenance and repair measures are not possible.

Tolerances and limits for light rail systems should be based on vehicle reactions, which in turn depends on the current track quality. This would permanently lead to a track condition which allows an operation that is comfortable for the passenger or economically profitable for the infrastructure manager.

The focus of this work is on the evaluation of the track geometry of LRT tracks, for which, based on the vehicle reaction, tolerance and limit values for maintenance measures are determined. Since data for sections in very bad condition were not available, artificially poor track conditions are generated by a successive increase of the signal amplitudes within the frequency range of the measured track geometries. The track measurement data is imported into a multi-body simulation software and the vehicle response is calculated on the passage of selected sections for a running time of five minutes. Then the vehicle reaction is evaluated on the basis of the passenger comfort, the track loading and the derailment coefficient, and the quality of the track position is determined. The load on the track body calculated by the multi-body simulation does not exceed permissible limits in comparison to the comfort value and the derailment coefficient.

To better illustrate the tolerances and limits for light rail systems derived from the vehicle reaction, a track geometry index is introduced. The Track Geometry Index allows a track manager to visualize the deterioration of the track layout and to divide it into different quality levels. The quality levels help infrastructure managers to determine the time for repair measures in a scheduled manner. Two LRT lines are examined in detail using the method developed here. The result indicates that the routes considered in this work were repaired too early.

Future work should detect irregularities in the road through continuously measured accelerations on the vehicle and in the track. In addition, acceleration data can then also be used for the assessment of longer sections of the route using the approach developed here.

1 Introduction

1.1 Motivation

The study seeks to promote Light Rail Transit (LRT) (Ger. Stadtbahn) as a viable transportation system alternative to the most expensive metro system (Verband Deutscher Verkehrsunternehmen 2014) and the lower capacity bus rapid transit systems. LRT is generally and informally defined as *"A railway system characterized by its ability to operate single or multiple car consists along exclusive rights-of-way at ground level, on aerial structures, underground on in streets, able to board and discharge passengers at station platforms or at street, track, or car-floor level and are normally electric, powered by overhead electrical wires"* (Parsons 2012). Although LRT systems offer much flexibility, they often encounter strong opposition (e.g. in Latin America) since often it is perceived as an unnecessary investment due to apparent higher capital costs (Gordon et al. 2001).

To promote LRT systems, and make them more attractive, it is necessary to reduce their total life cycle cost (LCC). A window of opportunity for cost reductions is within the maintenance and operation phase (Verband Deutscher Verkehrsunternehmen 2014), which represents 65 % of the total LCC (Kochs and Marx 2009) (see Figure 1).

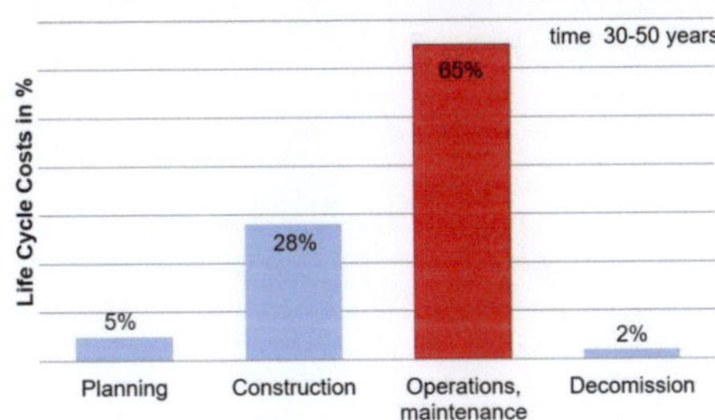

Figure 1:   **Share of costs for the Life Cycle of an Urban Railway (own work per (Kochs and Marx 2009))**

To help achieve this goal, maintenance should be performed in a timely and condition-knowledge-based manner. In addition, it should be properly coordinated with train operations to avoid disturbances or disruptions which incur in higher costs (Daniels 2008). However, knowledge-based maintenance in LRT systems is difficult since there are no established limits which could provide a systematic maintenance or renewal of the infrastructure (Parsons 2012). This is worsen by the fact that infrastructure managers follow maintenance practices based on empirical knowledge or conventional railway practices (Parsons 2012). The latter might not represent the most efficient approach since both systems display different behavior in terms of vehicle-rail interaction; causing different rates of track deterioration (Parsons 2012). It is important to develop an approach that helps establish maintenance/renewal intervention limits that can be used to conduct activities in a more cost-efficient manner.

1.2　Challenges

The creation of a structured and systematic approach for the maintenance of urban rail systems has been a recurrent topic of discussion and research (Kochs and Marx 2009). In Germany, an approach for data management and transferring has been developed for transportation companies. The approach identifies measures for the design, guidance and development of maintenance known as "Instandhaltungsmanagement" or maintenance management. The method consists of a basic strategy for the management of an asset based on the three basic constituents of railway infrastructure as shown in Figure 2.

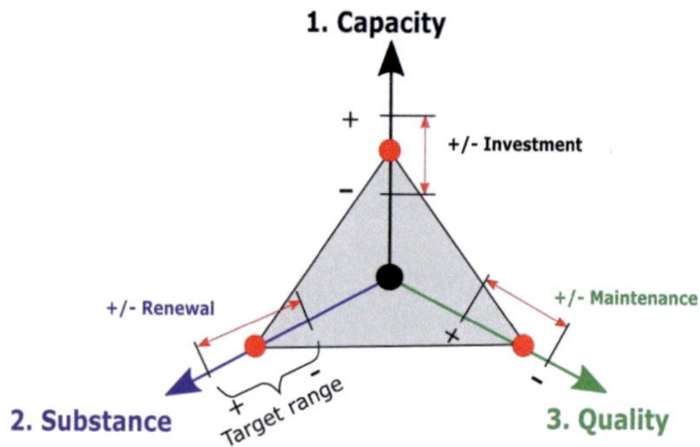

Figure 2:　Central aspects in the infrastructure investment strategy (own work per (Tzanakakis 2013))

The balance of these aspects represents the maximization of a system's efficiency (Tzanakakis 2013). To take care of the quality aspect, it is necessary to have in place an adequate maintenance policy. To achieve this goal, it is necessary to establish a process which includes periodic condition-based inspections deriving into meaningful track data for the development of knowledge-based maintenance and renewal. One of the biggest challenges to perform a knowledge-condition-based analysis lays on the availability of track geometry data and the openness of the asset management for co-operation. In praxis, an improvement of the maintenance policy would also require interdepartmental collaboration to achieve an optimum solution between operation and maintenance costs.

1.3 Objectives

The premise on which this research states that maintenance should be performed when needed and should be system specific, thus saving resources and reducing the overall costs of urban railway assets.

The research aims to develop an approach to establish limits of intervention for renewal of LRT tracks that consider track geometrical quality (condition of the track) and corresponding vehicle responses. To achieve this, the comfort level based on standard DIN 12299 (Deutsches Institut für Normung e. V. 2009), the track loading and derailment coefficient are evaluated using multi body simulations. This is done initially by using measured irregularities applied to the simulated track. Later, it is done through synthetic irregularities representing higher track irregularities (worst track condition or track quality).

Part of the objective is to present the limits in different ways, which might be helpful for easier visualization in practice, namely, power spectral density (PSD) curves, standard deviations, and a track geometry index (TGI). In specific, a TGI diagram could allow track / infrastructure maintenance managers to estimate the time span a given track deteriorates as well as the time at which the deterioration reaches alert, planned, corrective and immediate intervention / renewal limits.

1.4 Scope of the study

The study focuses in setting limits of intervention for LRT tracks based on available track data which are properly treated. The data is later used in multibody system (MBS) simulations of an LRT vehicle model based of the local transportation company in Stuttgart, Germany. Important assumptions are made which include the following:

- Tracks are subjected to the current basic maintenance regime (e.g. grinding, tamping, etc.), which might allow a temporary track geometry improvement.
- The vehicle used for simulation maintains an acceptable operational state ("nominal" condition) and undergoes no deterioration process.

- The track has been constructed through high quality standards and is in a good structural condition and presents no deterioration process.

1.5 Methodology

The workflow can be seen in Figure 3 which represent the higher level (level 0) of activities in this research. More detailed diagrams of the work are presented in Appendix XIII: Flow Charts for Formal Specifications.

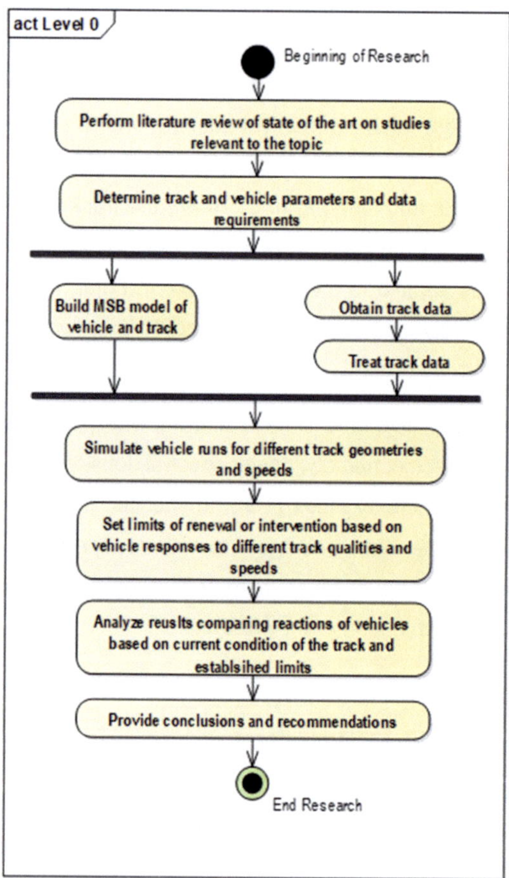

Figure 3: **Conceptual diagram (higher level) of the study (own work)**

1.6   Structure of the document

The thesis is divided in 6 chapters. Chapter 2 refers to track geometry quality, determining parameters and methods to establish it. It also provides reasons for the need to develop an approach to generate maintenance/renewal limits of intervention for LRT systems. Chapter 3 explores available data to perform the study. It includes methods to obtain the data, treat them for further analysis as well as methods to synthetically generate them. Chapter 4 deals with ballasted track construction, track maintenance and strategies, parameters to determine needs for maintenance and brief discussion of service life of track components. Chapter 5 deals with the determination and representation of track maintenance / renewal limits as well as an outlook to determine the influence that a suspension out of its nominal condition might exert in the determination of accelerations inside the vehicle's cabin. Chapter 6 provides conclusions and perspectives for future research, mainly regarding the determination of fault detection and identification of specific vehicles and suspension systems in relation to a given track geometry quality. In addition, further research would integrate track structural aspects and conditions, optimum maintenance regimes as well as reliability, availability, maintainability and safety (RAMS) and life cycle cost (LCC) analysis.

2 Track Geometry Quality

Track quality (TQ) relates to a specific system (e.g. High Speed, Mixed traffic, etc.). A distinction between systems becomes necessary since the type of traffic on a track influences the track geometry (Ferreira and Murray 1997). This is no exception to LRT systems, where construction and maintenance standards and methods, speeds and loads differ from other railway systems. To help determine the quality of LRT ballasted tracks in the future, this research proposes a working definition understanding that TQ in general depends on the geometrical and structural condition of the track (Xu et al. 2013), vehicle responses, comfort, RAMS and LCC.

Quality according to the International Organization for Standardization (ISO) (International Organization for Standardization 2015), is "the degree to which a set of inherent characteristics (i.e. distinguished features) fulfills requirements." Based on this definition, TQ can be defined as "the degree which a set of track characteristics (e.g. Parameters describing geometrical and structural conditions) fulfill specific requirements such as passenger comfort, RAMS and LCC" as shown in Figure 4.

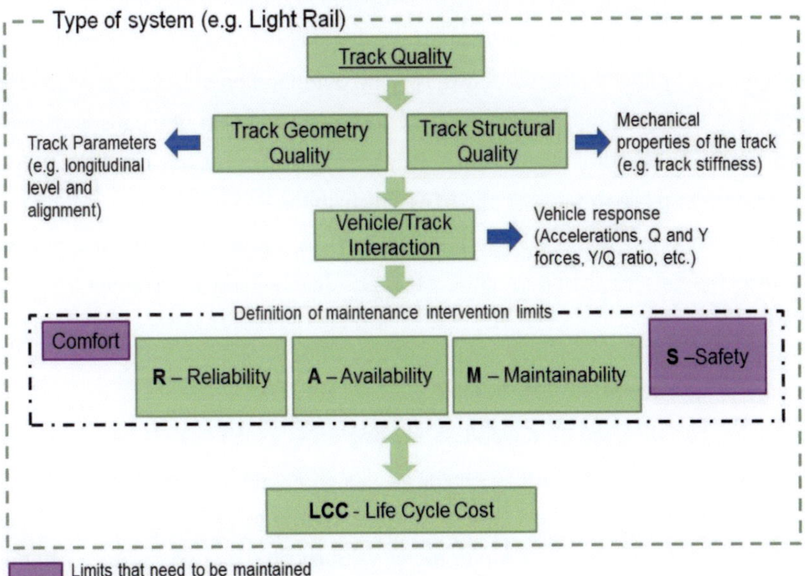

Figure 4: **Working Definition of Track Quality (own work)**

TQ needs to be evaluated through methods that determine the current state of a track and provide limits of intervention to restore its quality. TQ is commonly analyzed through its geometrical condition (Xu et al. 2013), usually referred as track geometry, which is described by a relative cartesian coordinate system centered to the track and measured clockwise (Deutsches Institut für Normung e. V. 2016).

2.1 Track quality deterioration and degradation

TQ deterioration is the process undergone by a track (Suiker 2004) under the passage of a large number of train axles that provoke plastic deformations in the granular sub-structure (Suiker 2004) causing track alignment deviations. As mentioned by (Xu et al. 2013), track deterioration depends on factors such as formation condition, in specific in relation to the effectiveness of the drainage system, ballast quality, traffic loading, type of track construction, dimension of track components, curvature, quality of materials, etc. The process of deterioration leads to track degradation (i.e. reduction of strength, efficacy, or value) in terms of safety, cost-efficiency of the operation and its renovation (Camacho et al. 2016). The deterioration of a track can be divided in three interrelated aspects: sub-structural, super structural and track geometry (Sadeghi and ASKARIAN 2007). Each of these lead to a degradation of their corresponding track subsystem while affecting the other two, i.e., the deterioration of the track geometrical aspect leads to the degradation of the track geometry; however, the causes of deterioration may lie on the structural characteristics / condition of the track (Xu et al. 2013). This agrees with (Haigermoser et al. 2014) which states that maintaining a good track geometry is important since it influences the dynamic behavior of vehicles, which in turn affects the deterioration rate of the track; possibly leading to an earlier track degradation.

2.2 Track quality deterioration rate

The deterioration rate can be defined as the degree at which the TQ decreases over time. The track deterioration can be geometrical or structural (Berawi et al. 2010), leading to a vehicle reaction which might differ form system to system and level of track degradation, i.e. changing of track parameter values within a short distance influence passenger comfort and operation safety. Furthermore, (Xu et al. 2013) explains that the deterioration rate of a track depends firstly on the initial quality (e.g. slower deterioration) and secondly on the current quality (Veit 2007), which if excessively decreased, leads to inefficient quality restoration.

2.3   Track geometry parameters

According to standard EN 13848-1 (Deutsches Institut für Normung e. V. 2016), there are five parameters that describe a track and its geometrical quality, namely gauge, vertical level, lateral alignment, cross level and twist (see Appendix I: Track Irregularities).

Modern railway systems measure the parameters as spatial signals through track recording vehicles (see section 3.1). In this work, track parameter deviations from their designed spatial position is known as track irregularity.

2.3.1   Track Irregularities

As expressed in (Iyengar and Jaiswal 1995), an irregularity is the variation from a long-term average level in the vertical and lateral profiles, which in general develops through sub-grade settlement, aging of the track components and elements and ballast loosening. Tracks are expected to always present irregularities due to the heterogeneous character of their components, type of operation, construction techniques and ground conditions (Iyengar and Jaiswal 1995; Garg and Dukkipati 1984). Track irregularities are important to consider in the vehicle-infrastructure interaction (Garg and Dukkipati 1984) since they cause vehicles to vibrate inducing forces on the track (Iyengar and Jaiswal 1995). Higher irregularities cause further and faster track deterioration (Xu et al. 2013) and damages to the vehicle (KNOTHE and GRASSIE 1993). Likewise, irregularities lead to accelerations in the passenger cabin influencing passenger comfort (Iyengar and Jaiswal 1995) and after long exposures, their health (Garg and Dukkipati 1984; Karis 2009). Furthermore, excessive irregularities for short wavelengths can in turn cause derailments (Garg and Dukkipati 1984).

The way to represent irregularities depends on the application (e.g. simulation) and available data (e.g. lateral irregularities for the left and right rails). Hence, there is the need for two coordinate systems, namely the rail coordinate system and track coordinate system. The rail coordinate system (Figure 5) references the positions of the rail along the x-axis of the track in relation to the designed track alignment. Where y_L and

z_L represent the left horizontal and vertical deviations of the left rail respectively and y_R and z_R represent the deviations of the right rail respectively.

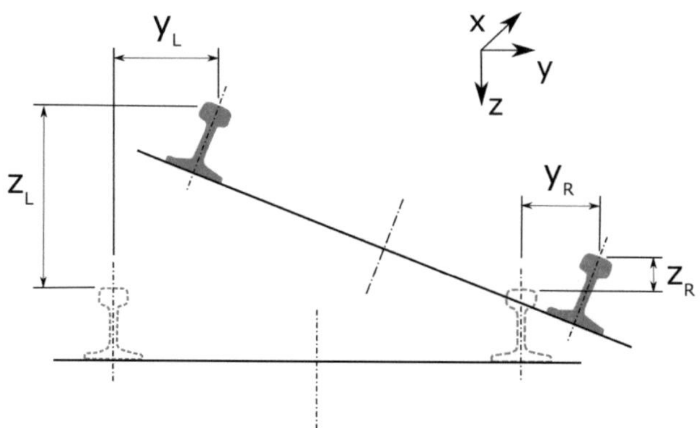

Figure 5: **Rail coordinate system (own work per (Andrea Haigermoser 2013))**

The track coordinate system (Figure 6) references the position of the rails in relation to the center line of the track. Where g and g_0 represent the measured and nominal gauge respectively, y_{CL} and z_{CL} represent the horizontal and vertical deviations with respect to the center line of the track and δ represents the angle measuring the grade difference between the left and right rail, determining the track's cross-level.

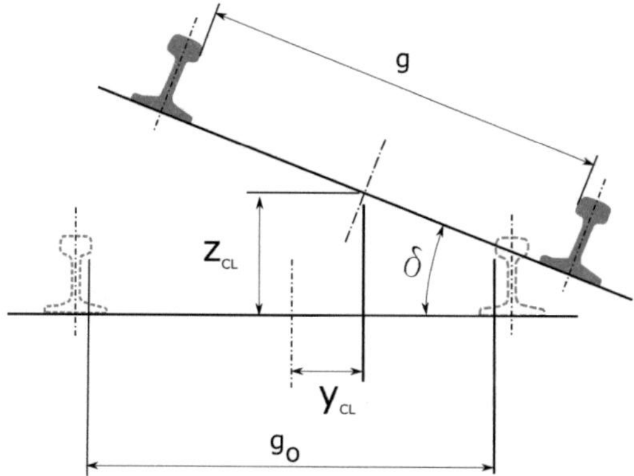

Figure 6: **Track coordinate system (own work per (Andrea Haigermoser 2013))**

The two coordinate systems are related through mathematical relations (see Appendix II: Conversion between Coordinate Systems). For example, the horizontal irregularity value (i.e. center line) used in this thesis is obtained from the average of the horizontal irregularity of the left and right rail:

$$y_{cl}(x) = \frac{y_L(x) + y_R(x)}{2} \tag{1}$$

To convert back to the track to rail coordinate system the following formulas are used:

$$y_L(x) = y(x) + \frac{g_o - g(x)}{2} \tag{2}$$

$$y_R(x) = y(x) - \frac{g_o - g(x)}{2} \tag{3}$$

where b_A represents the lateral spacing of rail-wheel contact points (i.e. 1,500 mm for standard gauge).

2.3.2   Track irregularities – wavelength

Wavelength is an important characteristic describing track geometry in relation to the vehicle/track interaction (Li et al. 2008). Standard EN 13848 stipulates three wavelength ranges for the assessment of track irregularities as shown in Table 1.

Wavelength band	Name
3m < λ ≤ 25m	D1
25m < λ ≤ 70m	D2
70m < λ ≤ 200m	D3 lateral (alignment)
70m < λ ≤ 150m	D3 longitudinal level

Table 1:   **Wavelength ranges for the assessment of longitudinal level and alignment per (Deutsches Institut für Normung e. V. 2016)**

Several reasons exist for grouping wavelengths. First, only certain wavelengths influence vehicle responses (e.g. accelerations in the vehicle's body) due to the vehicle's eigenfrequencies and travelling speed (Lewis 2011) through the following relationship:

$$\lambda = \frac{v_{veh}}{f_{veh}} \tag{4}$$

where λ is the wavelength of the track irregularity in [m], v_{veh} is the speed of the vehicle in [m/s] and f_{veh} is the eigenfrequency of the vehicle in [Hz].

Findings reported in (Andrea Haigermoser 2013) identified that D1 wavelengths are most effective for the description of the vehicle response in simulations. Another reason relates to the statistical evaluation of TQ. Long wavelengths display large amplitudes in comparison to shorter wavelengths (Andrea Haigermoser 2013). This means that longer wavelengths would dominate the SD values used in statistical evaluations.

Lastly, the choice of maintenance activities that have to be performed. For example, wavelengths to about 40 meters can be corrected by machine tamping or even other manual means when irregularities are isolated. Longer wavelength defects, which might not affect the response of the vehicle evaluated in this thesis, are usually corrected with more expensive and time consuming maintenance methods (Lewis 2011).

Non-affecting vehicle response wavelengths need to be filtered out form the track irregularity signals according to the speed and type of vehicle analyzed. In this research, the vehicle under study operates at 80 km/h maximum and has an eigenfrequency range of 1-2 Hz. Based on the vehicle characteristics, as shown in Figure 7, vertical or horizontal irregularities should be filtered between $3m < \lambda < 25m$ (i.e. D1 wavelength range).

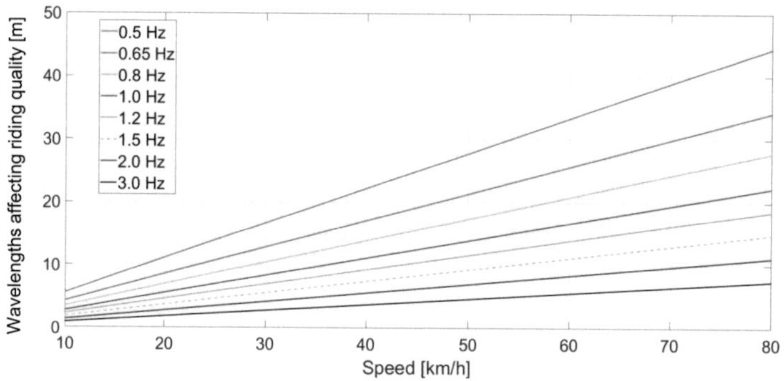

Figure 7:   **Wavelengths affecting riding quality (own work)**

2.4   European track geometry standards and limits of intervention

Track geometry quality (TGQ) is assessed through mainly two European standards, namely EN 13848 – part 1 (Characterization of track geometry) (Deutsches Institut für Normung e. V. 2008), part 5 (Geometric quality levels) (Deutsches Institut für Normung e. V. 2010) and part 6 (Characterization of track geometry) (The British Standards Institution 2014) and EN 14363 (Testing and simulation for the acceptance of running characteristics or railway vehicles) (The British Standards Institution 2016). In Germany a third standard from Deutsche Bahn AG (DB AG), RIL 821, sets limits of maintenance intervention for the German railway network based on track geometry measured by track geometry vehicles (DB Netz AG: Bautechnik, Leit-, Signal-, u. Telekommunikationstechnik 2004). Another standard, EN 12299 (The British Standards Institution 2009) (passenger ride comfort), mentions that comfort indexes could be used to determine track maintenance needs.

These standards are briefly discussed in the following subsections; setting the evaluation background of the study. Furthermore, the section includes a discussion about the intervention limits for LRT systems; setting arguments for the need to develop an approach that determines their maintenance limits.

2.4.1 Standard EN 13848–1

EN 13848–1 describes the minimum requirements for the analysis of track geometry parameters (Deutsches Institut für Normung e. V. 2016). The standard also describes measuring methods for the different parameters, namely inertial or versine (chord) based systems (see section 3.1). In addition, it establishes that if the chord method is used, the signal needs to undergo a decoloring process to eliminate distortions to amplitudes and phase introduced by the physical characteristics of the chord used (i.e. chord length and chord divisions) (Andrea Haigermoser 2013).

Amplitude distortions caused by a chord system are described through a transfer function whose magnitude lies between 0 and 2 for the different wavelengths contained in the signal (see Figure 8).

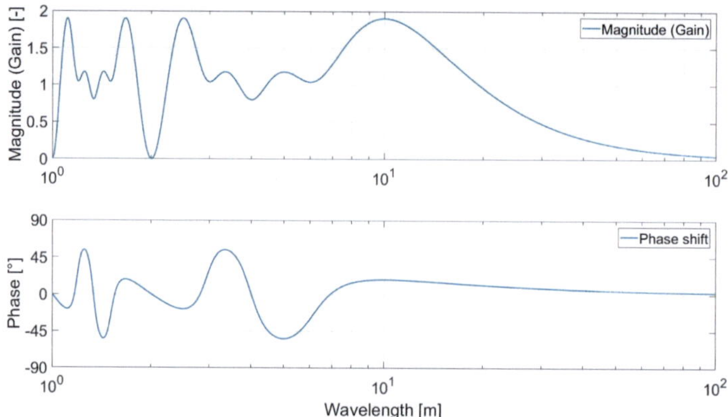

Figure 8:   **Example of transferring function of chord measuring (chord division 4m/6m) (own work per (Deutsches Institut für Normung e. V. 2016))**

It is important to realize that chord measuring systems have difficulties to provide reliable information about wavelengths greater than 70 meters.

To avoid gaps in the transfer function (i.e. values of 0), it is recommended that the divisions (i.e. a and b) of the chord total length L_{chord} are asymmetric (Andrea Haigermoser 2013) since as seen in Figure 9 symmetric divisions generate a gain of zero in the transfer function at several wavelengths (e.g. λ = 5 m).

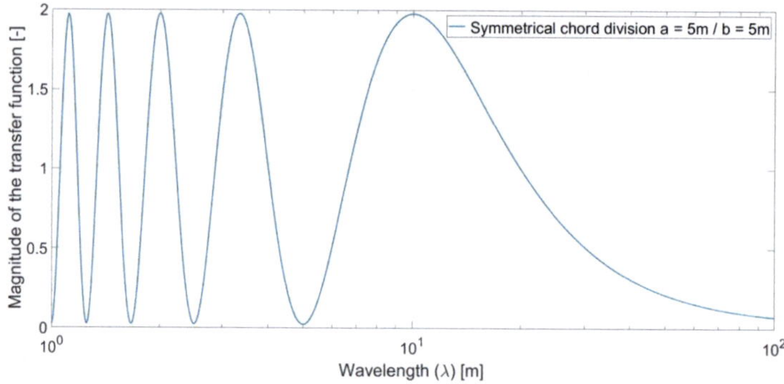

Figure 9:   **Example of symmetrical chord system (own work)**

The decoloring consists of the multiplication of the inverse of the transfer function of the chord system times the signal, which results in a new signal as seen in Figure 10.

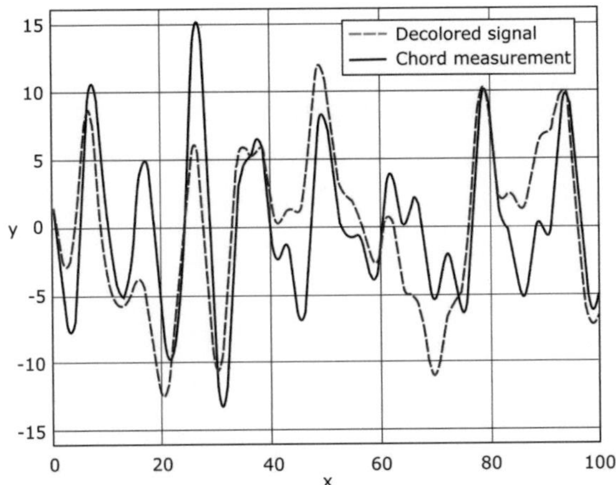

Figure 10: **Example of signal distortion due to chord measurement per (Deutsches Institut für Normung e. V. 2016)**

2.4.2 Standard EN 13848–5

EN 13848–5 deals with limits of intervention used for the assessment of the TGQ as shown in Table 2. Standard deviations or maximum values are provided as assessment quantities as discussed in section 2.5.1.1.

Alert Limit (AL)	The track geometry condition is analyzed and considered in the regularly planned maintenance operations
Intervention Limit (IL)	Level requiring a corrective maintenance, so an immediate action limit is not reached before the next inspection
Immediate Action Limit (IAL) "Safety Limit"	Measures to reduce the risk of derailment to an acceptable level need to be taken (e.g. line closure, speed reduction or track geometry correction)

Table 2: **Intervention Levels in EN 13848-5 (The British Standards Institution 2010)**

AL and IL limits do not consider vehicle/track interactions and values provided are purely indicative; i.e. meant for guiding maintenance policy (The British Standards Institution 2010). IAL is derived by experience and theoretical considerations accounting for wheel/rail interactions (The British Standards Institution 2010) and mainly concerned with safety; also important for the determination of LRT limits.

The track geometry evaluated through the standard accounts for discrete events (i.e. comparing to the signal's maximum values) and overall track conditions (i.e. comparing to the signal's standard deviations of normally 200-m segments). However, the standard acknowledges the need to include vehicle reactions in the determination of a more systematic maintenance policy by stating that maintenance might be directed either at upholding safety alone or additionally at achieving a good ride quality, lower life cycle cost or more attractive services.

Important for this work is to notice that the limit values presented have a lower boundary of 80 km/h (except for track gauge with a lower boundary of 40 km/h); making it not suitable for the detailed analysis of systems operating at lower speeds.

2.4.3   Standard EN 13848–6

Standard EN 13848–6 (The British Standards Institution 2014) establishes the characterization of TGQ. To do so, it determines that the methods used should be in accordance to the parameters established in part EN 13848–1 (see section 2.3) and that a TGI should be as simple as possible and reflect the principles of track/vehicle interaction. The standard describes methods for TGI assessments based on static evaluation (i.e. maximum values, standard deviation or combination of weighted standard deviations of parameters) or based on vehicle responses. The combination of standard deviations is evaluated and considered to be a better reflection of vehicle behavior than individual SDs (The British Standards Institution 2014). Dynamic methods are also presented in the standard. These methods are discussed in more detail in section 2.5.1.6.

The standard also mentions the possibility to use the Power Spectral Density (PSD) of signals, which is referred as a way that infrastructure managers could acquire better understanding and knowledge of the characteristics of irregularities, i.e. the wavelength range of track defects.

2.4.4   Standard EN 14363

In terms of track geometric quality, standard EN 14363:2016 provides three levels, namely QN1, QN2 and QN3. The latter represents the maximum absolute value in mm for different reference speeds. Table M.4 in the norm provides a value of QN3 for the alignment and longitudinal level for speeds below 160 km/h as shown in Table 3.

Reference speed [km/h]	Maximum absolute value (mean to peak) QN3 for wavelength range D1 [mm]	
	Alignment	Longitudinal level
$V_{veh} \leq 120$	13	16

Table 3:   **Track geometry limit values for QN3 level per (The British Standards Institution 2016)**

As it can be seen, the standard does not provide details for vehicles riding at a lower speed, which is the case of LRVs. In addition, as discussed in (The British Standards Institution 2014), QN3 values do not represent the most adverse maintenance; still allowing regular train operations.

A second aspect detailed in EN 14363 are running characteristics of a railway vehicle which are essential for the safe and economic operation of a railway system. The characteristics are vehicle speed, track condition (layout and geometry), and wheel/rail interface (The British Standards Institution 2016).

The standard establishes assessment values to determine the quality of the track in terms of the vehicle responses. The important values for this research are running safety in terms of the Y/Q (derailment coefficient) and track loading in the vertical direction Q, where Y is the lateral force on the track and the Q represents the vertical wheel force.

The standard establishes an evaluation process of the measured data and a minimum sampling rate of 200 Hz (The British Standards Institution 2016). For the evaluation, the standard establishes "test zones" covering varied track configurations. This work is concerned with test zone 1 which relates to straight and very large radius curves (i.e. > 600m). The signals are filtered to remove unwanted frequency contents and percentiles are calculated, then limit values are established to determine whether vehicle responses are acceptable (see Table 4).

Assessment value	Filter requirements	Percentile	Limit value
$(Y/Q)_{max}$	Low pass filter (20 Hz) Sliding mean method – 2-meter window and step length ≤ 0.5 meters	99.85%	0.8[1]
Q_{max}	Low pass filter 20 Hz	99.85%	Allowable pressure σ_z on subgrade see Appendix X: Vertical Stress Calculation

Table 4: **Conditions for processing the measuring signals and limit values per (The British Standards Institution 2016)**

2.4.5 Deutsche Bahn AG (DB AG) - RIL 821.1000 and RIL 821.2001

DB regulation establishes the principles and evaluation criteria for the inspection of the track geometry using the "German Railway's TRV".

The regulation establishes the condition of the infrastructure through four so-called disturbance/reaction (Ger. Störgröße / Reaktion - SR) assessment criteria that relate to the maximum relevant range up to which the track may be worn out without having to be renewed or replaced (Ger. Abnutzungsvorrat) as shown in Table 5:

---

[1] The maximum limit value of 1.20 shall be respected where 0.8 is exceeded according to table 4 in BS EN 14363:2016 [The British Standards Institution (2016)]. This number will be taken as the absolute maximum for safety (e.g. Immediate action – IAL required). 0.8 will be taken as the maximum value for normalization.

SRA	value above which scheduling of a repair is required from an economic point of view
SR100	value that includes the track's technical/economic range up to which the track may be worn out before renewal or replacement. If exceeded, a repair is required until the next regular inspection. The latest repair time is determined by the degree of overrun of the SR100 in connection with fault development
SRlim	value above which impairment of functionality is expected. A repair is to be carried out in the shortest possible time
Limit value	value at which the closure of the track and an immediate repair are required.

Table 5: **Disturbance / reaction categories per DB regulations per (DB Netz AG: Bautechnik, Leit-, Signal-, u. Telekommunikationstechnik 2004)**

The regulation provides values for different speeds per track irregularity as well as the way irregularities are measured. Table 6 shows the values for speeds lower than 80 km/h. As it can be seen the SRlim and Limit value are not provided.

				SRA	SR100	SRlim	Limit value
Parameter	Measuring basis [m]	Unit	Evaluation	< 80 [km/h]			
Longitudinal	2.6/6.0	[mm]	Zero/Peak	12	15	-	-
Twist	1.5 -19.5	‰	Zero/Peak	-	-	-	-
Cross level	-	[mm]	Average/Peak	10	13	-	-
Lateral	4.0/6.0	[mm]	Zero/Peak	12	15	-	-
Gauge	-	[mm]	1435/Peak	+15	+27	-	1430/1

Table 6: **Limit values at 80 km/h for the different assessment criteria per (DB Netz AG: Bautechnik, Leit-, Signal-, u. Telekommunikationstechnik 2004)**

2.4.6   Standard EN 12299 (comfort index)

Ride quality is expressed as the capability to operate a vehicle at any speed up to the maximum operating speed and at any passenger loading, from vibration and shocks to a specified level (Parsons 2012). In this study, ride quality is partially assessed through accelerations in the passenger cabin following the mean comfort evaluation as established in EN 12299, which takes into account the vibration exposure measured on the floor at different positions in the train (The British Standards Institution 2009). BS DIN12299 is specifically used for the analysis of comfort for passengers traveling in railway vehicle (including LRVs) for people in good health conditions. The methods presented in EN12299 can also however be used for measured or simulated vibrations.

In terms of its definition, comfort on a vehicle is a complex sensation produced during the application of oscillations and/or inertial forces, via the whole-body transmission caused by the movements of the vehicle's frame (The British Standards Institution 2009). Comfort is an important aspect of the operation quality of a railway service, but is difficult to assess since as mentioned in (Pollard and Simons 1984) comfort is judged on the expectations of a passenger based on the type of vehicle and service provided (e.g. high speed vs light rail).

The main quantity to measure vibration is the acceleration measured on a basicentric coordinate system at the point where the accelerations enter the human body (see Figure 11) considering the yaw, pitch and roll which correspond to the movement of the train around the coordinate axes.

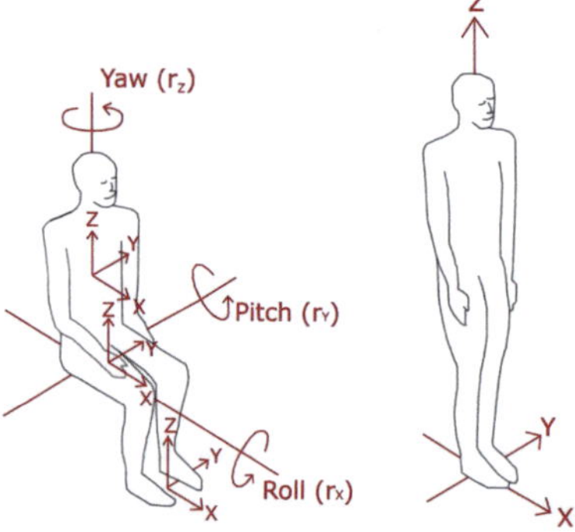

Figure 11:   **Basicentric axes of the human body (own work per (International Organization for Standardization 1997))**

There are several approaches to evaluate comfort in EN 12299. The first two are the simplified N_{MV} and complete N_{VD}, N_{VA} procedures for the evaluation of mean comfort level. The second is the continuous comfort for all axes C_{Cx}, C_{Cy}, C_{Cz}. Lastly, there is the evaluation for the comfort for discrete events P_{DE} and on transition curves P_{CT}. For the determination of the comfort, it is important to consider the actual state of the vehicle, for example, the nominal condition of the suspension, as well as the mass, center of gravity, etc. In that sense, the vehicle should be able to run under nominal service conditions. An approach to determine the nominal condition of the suspension, based on accelerations measured on the axle box and cabin, will be explored in chapter 5.

The mean comfort requires that accelerations are measured on the floor of the vehicle and is identified as suitable to determine the need to perform track or vehicle maintenance as shown in Table 7.

Comfort Index	N_{MV}	N_{VD}	N_{VA}
Passenger comfort	✓		✓
Vehicle assessment	✓		
Maintenance track	✓		
Maintenance vehicle	✓		
Motion quantities	Accelerations in all directions		
Measuring position	Floor	Floor	Floor /interfaces

Table 7:   **Aspects that might be evaluated through the mean comfort index per (The British Standards Institution 2009)**

To obtain the mean comfort value, accelerations are measured or simulated for five minutes which is the length of time the perception of what constitutes comfort for a representative group of people (The British Standards Institution 2009).

To provide more weight to frequencies at which the human body is more sensitive, the signals are filtered through bandpass weighted functions (Dassault Systems Simulia Corp. 2017) as shown in Figure 12.

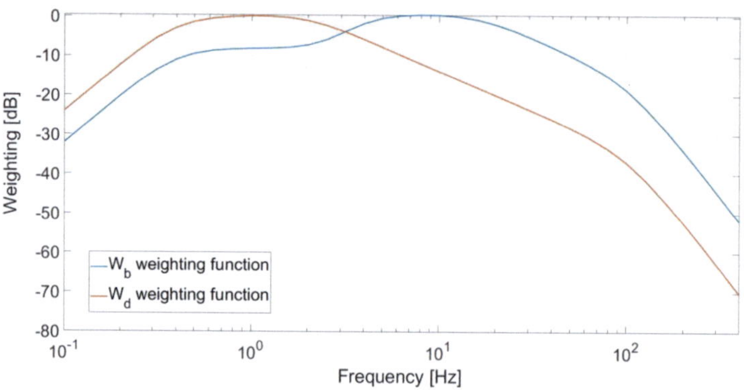

Figure 12: **Weighting curve W_b (z direction) and W_d (x,y direction) for mean comfort evaluation (own work per (The British Standards Institution 2009))**

The evaluation of the mean comfort index is taken at the vehicle's floor, the weighting function used for the evaluation is W_d for accelerations measured on the x and y directions. For the accelerations in the z direction the W_b weighting function is used as seen in the superscripts in Equation 5.

The filtered and weighted signals are grouped into five second periods and their root square mean (RMS) values calculated (resulting in continuous comfort). The 95th percentile of the RMS values are determined, and the mean comfort calculated with the following equation:

$$N_{MV} = 6\sqrt{(a_{XP95}^{W_d})^2 + (a_{YP95}^{W_d})^2 + (a_{ZP95}^{W_b})^2} \tag{5}$$

Appendix III: Mean Comfort Calculation provides an example for the calculation of the mean comfort index.

2.4.7   Intervention limits for Light Rail Systems

As mentioned in (Parsons 2012) intervention limits for the maintenance of LRT systems are not standardized. For example, in the United States (USA), limits are unique to each transit property based on their needs and objectives. The values of the standards would inherently expect that a certain amount of deterioration will occur, but that within maintenance cycles the riding quality would not be impaired.

In Germany, an innovative maintenance management approach has been detailed in a document created by the association of public transportation companies (Kochs and

Marx 2009). The approach proposes four track failure categories that match the levels expressed in EN 13848–1 (see section 2.4.2), the technical specifications for the interoperability of trans European high-speed trains and the German Railways standard under the RIL 821 regulation (DB Netz AG: Bautechnik, Leit-, Signal-, u. Telekommunikationstechnik 2004). Figure 13 depicts the proposed intervention levels for the maintenance of light rail systems in Germany.

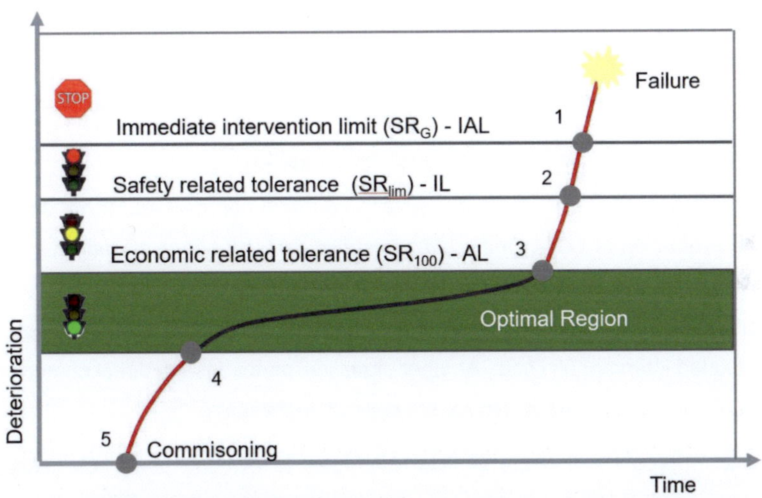

Figure 13:   Schematics of maintenance intervention levels conceived for LRT systems (own work per (Kochs and Marx 2009))

However, the document states that values for the limits need to be developed and only limits for the gauge and components of turnouts are presented. Likewise, SSB (Stuttgarter Straßenbahnen AG – SSB AG), which provided the data for this study, has established some assessment values, but are not used in practice (Benz 2016). In addition, the values established (see Table 8) do not hold a relationship to vehicle responses. Therefore, there is a need to develop values for the evaluation of the TGQ.

Nr.	Parameter	Unit	SR_{100}	SR_{lim}	SR_g
1	Gauge	mm	± 10	± 15	± 20
2	Lateral (tangent)	mm	n.d.	n.d.	n.d.
3	Lateral (curve)	mm	n.d.	n.d.	n.d.
4	Cross level	mm	n.d.	± 36	n.d.
5	Vertical	-	n.d.	1:300	1:150
6	Twist	-	n.d.	1:300	1:150

Table 8: **SSB's specified limits of intervention for measured parameters per (Benz 2016)**

2.5 Methods to determine track geometry quality

In regular railway systems, there are several common approaches for the determination and evaluation of TGQ. One relates to statistical evaluations of track irregularities using weighted or simple standard deviations of the given parameters to create track geometry or quality indices. The results are compared to standard values to determine safety related intervention limits. Another statistical method is the direct comparison of maximum values of the track irregularity against limit values.

Additional approaches include vehicle responses to evaluate the track geometrical quality in relation to riding quality and safety or the analysis of signals in the frequency domain through the conversion of the spatial domain signals to their power spectral density or other frequency domain methods (i.e. wavelet analysis). In case of the PSD, the curves obtained are compared to standard PSD curves. The following subsections explore some of the most important approaches to determine the TGQ around the world.

2.5.1 Based on track irregularities – statistical methods

The need to formulate an standard track quality index had already been mentioned in (Fazio and Corbin 1986) back 1986. Today it is widely accepted that track quality is measured by the deviation of the track geometry parameters from their designed values. Statistical methods are used to determine the geometrical quality of a track based on data measured through track recording vehicles (TRV). The information provided though evaluations help infrastructure managers determine maintenance requirements and prevent accidents. Several countries have developed statistical methods. In the United States, one method uses the summation of the squares of deviations to create a roughness index (Sadeghi 2010). Another US method looks at the calculated length of irregularities and obtains a relationship to the theoretical length of the track for a given segment. The method sought to overcome problems of compliance of other methods with FRA regulations (Zhang et al. 2004) and may be used to evaluate vehicle and track interaction by including vehicle characteristics (Lee 2005). Likewise, TQIs might be used to evaluate the effectiveness of track maintenance activities (Lee 2005).

This subsection shows a compilation of common statistical methods used in the determination of TGQ.

2.5.1.1   Standard deviation and maximum values methods

These correspond to methods that directly compare standard deviations or maximum values of measured parameters with prescribed limits. One of such methods is the standard deviation index (SD Index), which considers seven standard deviations corresponding to measured values on a track segment. The SD is calculated with the following formula:

$$\sigma_i = \sqrt{\frac{1}{n}\sum_{j=1}^{n}\left(x_{ij}^2 - \overline{x}_i^2\right)} \tag{6}$$

$$\text{with} \qquad \overline{x} = \sum_{j=1}^{n}\frac{x_{ij}}{n}$$

where σ is the standard deviation and x_{ij} is the measurement value in mm at the jth sampling point of a track segment and n is the number of points in the track segment.

Another method is prescribed in European standard EN 13848 – 5 which establishes a direct comparison to limit values for the vertical, horizontal and gauge irregularities at different speeds and for wavelength band D1 as shown in Table 9. The standard deviations are calculated for track segments of 200 meters for the longitudinal level and alignment, and for 100-meter segments for gauge (Berawi et al. 2010).

Speed	D1 wavelength band in [mm]	
	Profile	Alignment
V ≤ 80	2.3 – 3.0	1.5 – 1.8
80 < V ≤ 120	1.8 - 2.7	1.2 – 1.5
120 < V ≤ 160	1.0 – 2.4	1.0 – 1.3
160 < V ≤ 220	1.2 – 1.9	0.8 – 1.1
220 < V ≤ 300	1.0 – 1.5	0.7 – 1.0

Table 9: **SD limit values for profile and alignment parameters for Alert Limit (AL) per (Deutsches Institut für Normung e. V. 2010)**

Values in the standard can also be compared directly to wave amplitude limits for D1 and D2 wavelength bands and intervention limits described in section 2.4.2. Worth noticing is that the standard determinates that for speeds less than or equal to 40 km/h, the AL and IL limits can be relaxed (Deutsches Institut für Normung e. V. 2010).

EN 14363 determines as well standard quantities for the assessment of track geometrical quality for both standard deviations and maximum values for the alignment and vertical irregularities. Three quality levels are defined for different speed bands, namely Q1 to Q3. Q1 refers to the level in which maintenance is required within the level of normal/scheduled maintenance. Q2 required short term maintenance and Q3 represents a section with usual track geometrical quality. Signals measured require filtering for different wavelength band depending on the study.

2.5.1.2   Track Geometry Index (TGI)

Indian Track Geometry Index represents a synthetic value based on the average of weighted indices of the different geometry parameters (Liu et al. 2015), namely unevenness index (UI), alignment index (AI), twist index (TI) and gauge index (GI). The indices are the result of exponential functions that assess the condition of the parameters with respect to the standard deviations of 200-meter track segments (Berawi et al. 2010) for different states and conditions of the infrastructure ($SD_{measured}$) (Camacho et al. 2016) in a range defined by the standard deviation of a newly laid track (SD_{new}) and the standard deviation (SD) of a track needing urgent maintenance ($SD_{maintenance}$) (Research Design and Standards Organisation 1997). The parameter indices are calculated with the following mathematical relationship:

$$TI\backslash UI\backslash AI\backslash GI = 100 * e^{\frac{SDmeasured - SDnew}{SDmaintenance - SDnew}} \qquad (7)$$

Different weights are assigned to each parameter to consider their effect on an Indian Ride Index (Research Design and Standards Organisation 1997) as shown in the following equation:

$$TGI = \frac{TI + GI + 2UI + 6AI}{10} \tag{8}$$

The TGI was developed to overcome limitations displayed by an earlier method called composite track record (CTR) used to assess TGQ. These deficiencies were: lack of correlation of each parameter with riding quality, high sensitivity of the composite track record to minor changes in track geometry, the contribution of each parameter to riding quality and the consideration of only exceedance values instead of a continuous quality (Research Design and Standards Organisation 1997).

The limit values established in the TGI standard are presented in Table 10.

Parameter	Chord length [m]	SD - Newly laid Track [mm]	SD - Urgent Maintenance [mm]	
			Speed < 105 km/h	Speed > 105 km/h
Unevenness	9.6	2.50	6.20	7.20
Twist	3.6	1.75	3.80	4.20
Gauge	1.0	1.00	3.60	3.60
Alignment	7.2	1.50	3.00	3.00

Table 10: SD values in mm for newly laid track and track needing urgent maintenance per (Research Design and Standards Organisation 1997)

A classification of the limit values that determine the need for maintenance are presented in Table 11.

TGI Value	Maintenance requirement
TGI > 80	No maintenance required
50 < TGI > 80	Need basic maintenance
36 < TGI < 50	Planned maintenance
TGI < 36	Urgent maintenance

Table 11: Classification of maintenance with TGI per (Talukdar et al. 2006)

According to (Sadeghi and Askarinejad 2010) the TGI is a practical and reliable method for the determination of TGQ. It provides information on the track quality for a continuous length of track rather than highlighting isolated low-condition locations, it gives weightings to different parameters according to their effects on a ride index and it is not overly affected by minor changes in the track geometry (Sadeghi and Askarinejad 2010).

2.5.1.3   J-synthetic Coefficient

Known as J index, the method was developed in Poland to determine track geometry conditions. The coefficient is calculated with the standard deviations of the horizontal, twist, horizontal and gauge parameters measured with a 10-meter long chord (Sadeghi 2010).

The synthetic TQ coefficient specifies allowable deviations as shown in Table 12.

Speed [km/h]	30	40	90	120	160	200
J index	12.0	11.0	6.2	4.0	2.0	1.4

Table 12:   **J Index allowable values at different speeds per (Sadeghi 2010)**

As it is expected, the J index can be higher at lower speeds. This fact is of interest to the determination of light rail track geometry indexes.

2.5.1.4   Five parameters of defectiveness (W_5)

The W_5 method is another index developed in Poland to determine the quality of tracks (Chudzikiewicz et al. 2017). The method treats the defectiveness of each parameter, gauge, vertical level, horizontal level, cant, twist and gauge, as an independent event.

The coefficient of parameter defectiveness W for each parameter, is calculated as the ratio of the sum of the number of assessment sections which exceed an allowable value of defectiveness and the total number of sections. Irregularities are measured with an 18.6-m long chord.

As explained in (Chudzikiewicz et al. 2017), the W_5 method defectiveness is evaluated against maximum allowed values which determine the need for maintenance as shown in Table 13.

Evaluation of line	New	Good condition	Sufficient condition	Indicating insufficient condition
Value W_5	< 0.1	< 0.2	< 0.6	> 0.6

Table 13: **Quality qualifications of track lines for the five parameter defectiveness (W5) per (Chudzikiewicz et al. 2017)**

As it can be observed the higher the value of W_5 the more defective the track.

2.5.1.5 Fractal Analysis

The fractal analysis has been explored in the United States to characterize the roughness of track geometry as an indicator of track geometry condition (Hyslip 2002). Fractal analysis provides a good indicator of the geometrical signature of the track through numerical values (i.e. dimensions) while accounting for the magnitude of roughness variations and their frequency (Hyslip 2002) which surpasses other methods such as the roughness index.

Fractal analysis is able to directly compare the roughness characteristics of different length of tracks since the measurement is independent of length (Hyslip 2002). This allows to rank and categorize the track. It also allows for the determination of the so called "equivalent utility", which refers to the dynamic responses of vehicles (Hyslip 2002). For example, two sections of track will have an equivalent utility if their roughness are similar with respect to dynamic ride performance (Hyslip 2002).

The fractal analysis in (Hyslip 2002) was carried out through the "divider" or "ruler" method (Andrea Haigermoser 2013). Track geometry displays a bi-fractal aspect represented by two distinct linear portions of the data (Andrea Haigermoser 2013) which can be approximated with two straight lines with different fractal dimensions (D_{R1} and D_{R2}) and a breakpoint (change in slope) as seen in Figure 14.

The 1st order fractal dimension is associated with the overall shape of the line (i.e. larger wavelengths). The 2nd order is associated with the texture of the line (i.e. shorter wavelengths). The slope of the two lines as well as the location of the fractal breakpoint provide numerical parameters for the characterization of the roughness; in specific, indicates the point of separation between the relatively small and large scales (Hyslip 2002).

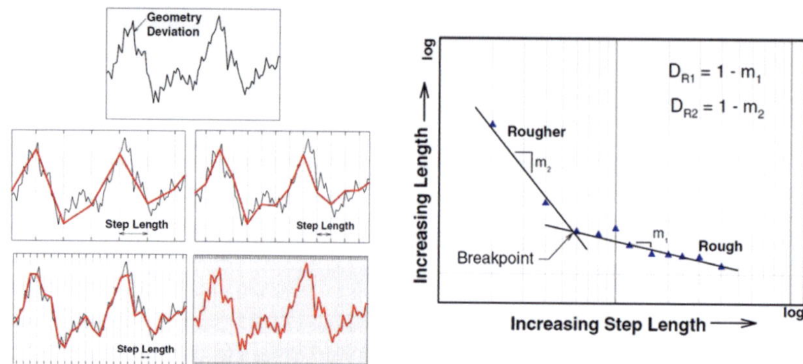

Figure 14:   **Fractal plot example pattern of geometry deviation for the track geometry shown per (Hyslip 2002)**

Fractal analysis provides useful numerical values that can be used to correlate tack geometry and track structural condition (Andrea Haigermoser 2013) (Hyslip 2002).

2.5.1.6   MDZ – A Number

The MDZ method of the Austrian Federal Railways (ÖBB) involves the vehicle movement in the determination of track (geometrical) quality. The method analyzes the movement of the center of gravity of the car body in the vertical $\Delta z'$ and lateral $\Delta y'$ directions within a given length L_A of the track. The method is based on an over simplified rigid model with the center of gravity located at 1.5 meters form the top of the rail (Luber 2011). Effects of springs and dampers are accounted as a function of velocity by introducing a constant value (Luber 2011).

With this method different parameters can be combined under one index which can be used to determine changes of the track over time, assisting infrastructure managers to develop maintenance strategies and intervention measures. To determine a change in quality over time (i.e. deterioration), the following relationship is used:

$$Q = Q_0 * e^{-bt} \tag{9}$$

where Q is the track quality over time represented by MDZ_a, Q_0 is the initial quality of the track (quality of track at the end of construction), b represents a rate of deterioration and t represents time or tonnage driven on the track.

As mentioned in (Veit 2007), it has been determined that the deterioration rate of the track depends on its current quality and the description of track geometrical quality should always be described by the quality and a rate of change. In addition, a very important conclusion is that the initial track quality cannot be reproduced (never mind improved) through maintenance. In this sense, the initial track quality is very important (achieved through higher capital costs). Lastly, separating capital works and a renewal strategy is inappropriate since optimization can only be achieved with a single comprehensive strategy.

2.5.2 Vehicle response analysis (VRA) methods

Track inspection methods based on statistical approaches present deficiencies in the determination of track quality since they do not relate to vehicle responses. In 1986 was already recognized that vehicle responses should be included in the determination of performance-based TQIs (Fazio and Corbin 1986). This section presents some methods based on vehicle/track interaction which seek to determine TQIs by correlating track parameters and vehicle responses. The values are obtained from synthetic or measured signals for the track irregularities and multibody simulations and/or test for the vehicle responses.

2.5.2.1 VRA-Method in the Netherlands

The VRA method in the Netherlands relates track irregularities to vehicle responses through linear transfer functions evaluated in the frequency domain. Three vehicle types are simulated in MBS at five different speed ranging between 40 – 160 km/h. In total there are 15 scenarios. The track parameters evaluated are longitudinal level, alignment (lateral irregularities), cant and gauge. The vehicle responses consist of passenger comfort calculated from horizontal accelerations a_y and vertical acceleration a_z. It also consists of horizontal forces Y and vertical forces Q. The combination of all parameters results in 240 transfer functions.

The first step in the implementation of the method is to transform the signals to frequency domain for 600-m track segments. The system is considered a multiple input multiple output (MIMO) so it is possible to calculate the vehicle response to each input signal by a multiplication with the transfer function (Andrea Haigermoser 2013). The resulting outputs are converted back to a spatial domain and added (i.e. by superposition) (Andrea Haigermoser 2013).

2.5.2.2   WGB-Method

WGB is a multibody numerical simulation VRA method developed by the German Railways (DB AG) to assess TGQ. It includes the interaction of track parameters and vehicle responses evaluated against maximum values (e.g. per EN 14363). The relationship between the track and vehicle responses is calculated through assessment functions produced through a regression analysis of the vehicle/track interaction (Andrea Haigermoser 2013).

The vehicle response parameters are obtained though simulation of validated multibody simulation (MBS) vehicles. The parameters are the sum of lateral forces per wheelset $\sum Y$, the quotient of lateral and vertical forces per wheel Y/Q, the maximum vertical wheel force Q_{max}, the minimum vertical wheel force Q_{min}, the maximum lateral car body acceleration a_y, the maximum vertical car body acceleration a_z (The British Standards Institution 2014).

The track parameters are longitudinal level (LL), lateral irregularities (AL) and cross level (CL) which are assessed sequentially taking the superposition of lateral and vertical track defects into account (The British Standards Institution 2014). The input parameters are amplitudes (amp) and gradients (gr) of the separated single defects as well as local permitted speed (V) and local curvature (cr) of the track (The British Standards Institution 2014).

The output values can be analyzed statistically to determine TQI values for a given track section depending on the selected level of aggregation (The British Standards Institution 2014).

2.5.2.3   Transfer Functions (TF)

A similar approach includes system identification to characterize vehicle responses to be able to predict the effects of track geometry on vertical vibration and track loading. The approaches are based on wavelengths of measured or synthetically generated track irregularities and vehicle reactions that result from testing or simulations (e.g. accelerations measured at the axle box). An approach making use of system identification methods is the Typical Transfer Function method (TTF-method) developed in Austria. The method uses multibody simulations along with the system identification to create the vehicle/track transfer functions capable of assessing track geometry with respect to vehicle response (Andrea Haigermoser 2013).

(Luber 2011) describes a track geometry assessment method (TGA) based on empirical transfer functions evaluated in the frequency domain which describe the track irregularities and the corresponding vehicle responses for representative vehicles. The method is based on the prediction of the vehicle's forces caused by track irregularities (Luber et al. 2010). The predictions are evaluated according to standard EN 14363. The transfer functions are calculated through system identification methods (Luber et al. 2010). The irregularities considered are longitudinal level (z), lateral (y) and cross level (δ). The vehicle responses considered are the sum of lateral loads (Σy) and maximum vertical loads (Q) for the vehicle response. The combination of the parameters resulted in six empirical transfer functions. Since the subsystems determined are single-input-single-output (SISO) time invariant linear systems, the outputs can be added according to the superposition principle (Luber et al. 2010) (see Figure 15). The resulting vehicle response force is a function of all geometrical parameters; representing a multiple-input-multiple-output (MIMO) system (Luber et al. 2010). The method was developed using multibody simulations on Siemen's vehicle models.

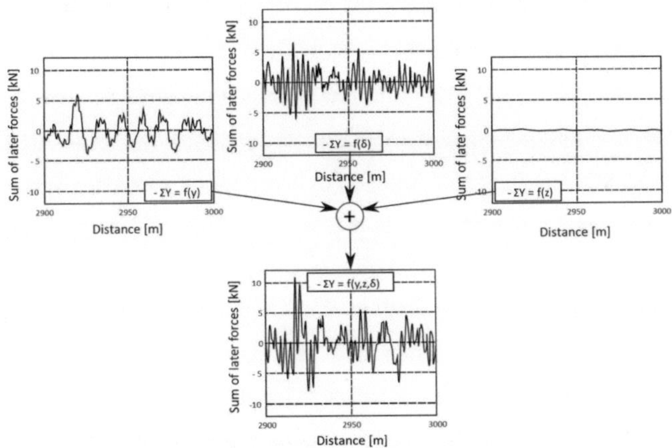

Figure 15: **Principle of superposition in LIT systems ΣY$_{MISO}$ from individual ΣY$_{SISO}$ per (Luber et al. 2010)**

2.5.2.4 Performance-Based Track Geometry (PBTG)

Track geometry inspection method developed in the USA based on neural networks (Andrea Haigermoser 2013). The method seeks to relate, in real time, complex patterns and nonlinear relationship of inputs and outputs such as the relation of track geometry with vehicle responses for the three most sensitive vehicles in the North American context. The central aspect of the PBTG is a group of neural networks trained from actual track geometry and vehicle response test results (Li et al. 2006). The main inputs are curvature, super elevation (long wavelengths), cross level (short wavelengths), vertical surface of both rails, lateral alignment of both rails, gauge and vehicle operational speed. The outputs are vehicle response parameters, in their studies, vertical (Q) and lateral (Y) forces and the derailment coefficient (Y/Q) (Li et al. 2006).

The motivation to develop the method resulted from observing that many times track geometry not exceeding current limits caused poor or adverse vehicle responses. Likewise, areas that show an exceedance of limits, not always caused poor vehicle reactions (Li et al. 2006). The method seeks to help railways to find and correct track segments that may produce poor vehicle performance (e.g. mainly regarding safety) and potentially reduce derailment incidents because of poor track geometry (Li et al. 2006). The method can also be used to identify vehicles that begin to deteriorate (Andrea Haigermoser 2013). On board accelerometers can be used to identify locations in need of maintenance.

2.5.2.5 Axle box accelerator performed based indicators

More recently, the determination of a track quality indicator was presented in (Chudzikiewicz et al. 2017). The indicator is based on track monitoring consisting of acceleration signals obtained from inertial sensors placed on the axle box of a vehicle in operation (Chudzikiewicz et al. 2017). The data is treated accordingly (e.g. filtered) and its power spectral density used in a mathematical expression analogous to the comfort index which includes the effects of velocity. An important aspect was the use of GPS and map matching algorithm which localize the identified faults on the track (Chudzikiewicz et al. 2017). A correlation of coefficients between the track quality indicator calculated and the standard deviation track geometry indices W5 and J-coefficients was performed; demonstrating a strong positive relation between 0.79 and 0.96. between the statistical based coefficients and the perform-based indicator.

2.5.3 Frequency analysis

Many signals are characterized as random since their variation outside an observed interval can only be specified in statistical terms of averages (Stoica and Moses 2005). Signals are most commonly obtained by temporal or spatial sampling of continuous signals. In previous sections methods to establish a TGQ from spatial domain signals were discussed. However, an assessment of track geometry is not easily done with signals in the spatial domain, e.g. it is difficult to determine the shape of track irregularities and their wavelength content (Andrea Haigermoser 2013), obtained easier through a Fourier transform.

The Fourier transform takes a periodic function in time (or space) and decomposes it into the sum of a constant and a series (sum) of sines and cosines (sinusoids) consisting of different frequencies as shown in the following expression.

$$f(t) = \frac{1}{2} a_0 \sum_{k=-\infty}^{\infty} (a_k cos 2\pi kt + b_k sin 2\pi kt) \tag{10}$$

where k represents the frequency, t represents the time, a and b represent coefficients of each frequency.

Fourier transformations have been performed with the MATLAB© function fft() representing the Fast Fourier Transform (FFT) which is a method that increases the computation efficiency. However, signals present an issue called leakage when analyzed in the frequency domain. This can be mitigated by using window functions. According to (Harris 1978) windows are weighting functions applied to reduce the spectral leakage associated with finite observations. In specific, spectral leakage refers to a spread of peaks in a spectrum causing them not to appear as sharp as they should be (Cimbala 2007). The latter occurs since signals in time/spatial domain do not always start and end at the same phase in the cycle (Cimbala 2007). Hence, as explained by (Harris 1978), windowing is applied to reduce the order of discontinuity at the boundary of the periodic extension; achieved by matching as many orders of derivatives as possible at the boundaries. This is achieved by setting the derivatives to zero or near to zero at the boundaries as it can be seen in Figure 16.

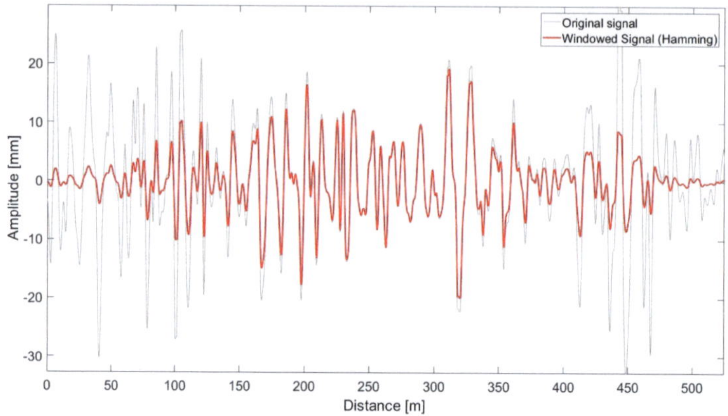

Figure 16: **Hamming window applied to signal to reduce spectral leakage (own work)**

Yet, as discussed in (Cimbala 2007), windowing causes a loss of information, leading to an overall reduction in amplitude of the frequency spectrum. To compensate this loss, the amplitude of the FFT must be multiplied by a factor of $\sqrt{8/3}$. In Figure 17 the factor has been applied and as it can be observed, the magnitude difference between the original signal (in dashed grey) and the windowed signal (in red) has been reduced (in green). However, it must be mentioned that leakage cannot in general be eliminated.

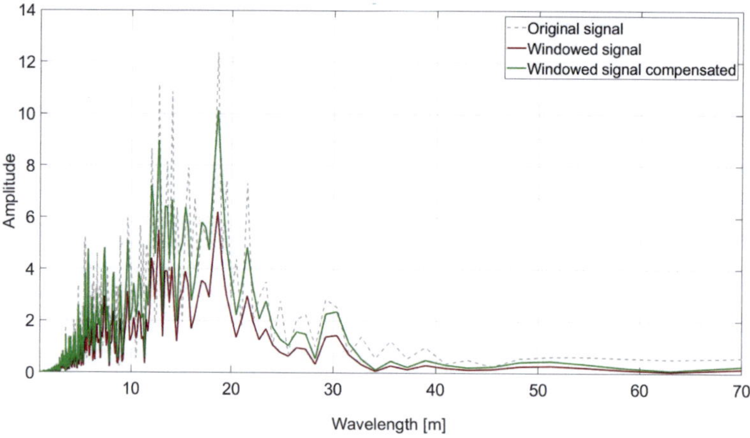

Figure 17: **FFT of signals depicting amplitude's differences (own work)**

2.5.3.1 Power Spectral Density

A representation of the frequency content of a signal is the power spectral density. According to (Barry Van Veen 2013), the power spectral density, characterizes a stationary random process in the frequency domain which is the result of the autocorrelation function $R(\tau)$ of a time (or spatial) domain $x(t)$ signal (Dassault Systems Simulia Corp. 2017) or simply put, the convolution of the signal with its complex conjugate.

$$R(\tau) = x(\tau) * x(-\tau) = \int_{-\infty}^{\infty} x(t)e^{-j\omega t}\, dt \tag{11}$$

where $R(\tau)$ denotes the autocorrelation function and τ denotes the lag of the signal.

A common expression to calculate the PSD is presented in (The British Standards Institution 2014)

$$S_{xx}(\omega) = X^*(\omega)X(\omega) \tag{12}$$

The Fourier transform is commonly used to calculate the PSD. However, this method produces a noisy signal (Andrea Haigermoser 2013) and requires windowing and compensation to mitigate the leakage effect. For smoother signals the pwelch method is used (Andrea Haigermoser 2013). The pwelch function returns the PSD of an input signal using the average of overlap estimators and can be calculated with the MATLAB function *pwelch()* (Mathworks 2018a). The pwelch function uses as default eight segments and a 50% overlap. Each segment is windowed with a Hamming window, modifying the periodogram. The modified periodograms are then averaged to obtain the PSD estimate. Since the segments usually overlap, data values that are tapered at the beginning and at the end of each segment, occur away from the ends of adjacent segments. This guards against the loss of information (i.e. leakage) caused by windowing (Mathworks 2018a) so no compensation factor needs to be applied. Smoothing of the PSD can be achieved with an increase of number of segments as it can be seen in Figure 18. Increasing the number of segments needs to be done with care since signals can deviate from their shape as the number of segments increase. This is especially true for signals without bandwidth limits such as cross level and gauge (Andrea Haigermoser 2013).

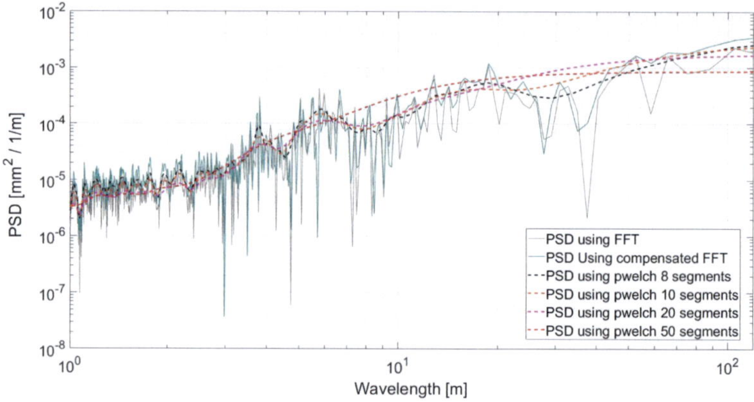

Figure 18: **PSD smoothing using pwelch function with different number of segments (own work)**

In track quality determination, PSD can help characterize the overall track geometry of a section or complete line as mentioned in standard EN 13848-6 (The British Standards Institution 2014). The characterization is in terms of wavelength and amplitude (Berawi 2003) that reveal typical peaks of repetitive track failures or irregularities. The irregularities normally correspond to the length of the rail and joint position or other periodicities in the support system of the track (Andrea Haigermoser 2013) (e.g. spacing of sleepers as shown in (Liu 2015)) contained in the spatial domain signal (The British Standards Institution 2014). PSDs are presented in an spectrogram that describes the change of spectral density function with respect to frequency (Lei 2017) where the higher a curve, the higher the track irregularities.

Several countries have established PSD standards for the analysis of their tracks, which can also be used to generate synthetic signals used in numerical analysis. The standards are based on analytical expressions of the PSD function (Berawi 2003) for different levels of track geometry irregularities. The most used standards in research are the European ERRI B176 standard and the FRA used in the United States.

The ERRI B176, sometimes identified as the German PSD, is a model characterized by one-sided spectrum. It is represented by analytical formulas for two categories: low and high irregularities. The empirical expressions to determine the PSD curves are presented below and the parameter values are presented in Table 14 (Lei 2017).

PSD vertical irregularities

$$S_v(\omega) = \frac{A_v * \omega_c^2}{(\omega^2 + \omega_Y^2) * (\omega^2 + \omega_s^2)} \qquad [\text{m}^2 / \text{rad/m})] \qquad (13)$$

PSD lateral irregularities

$$S_a(\omega) = \frac{A_a * \omega_c^2}{(\omega^2 + \omega_Y^2) * (\omega^2 + \omega_s^2)} \qquad [\text{m}^2 / \text{rad/m})] \qquad (14)$$

PSD cross level

$$S_{cl}(\omega) = \frac{(A_v * \omega_c^2) * \omega^2}{(\omega^2 + \omega_Y^2) * (\omega^2 + \omega_c^2) * (\omega^2 + \omega_s^2)} \qquad [\text{m}^2 / \text{rad/m})] \qquad (15)$$

PSD gauge

$$S_g(\omega) = \frac{(A_g * \omega_c^2) * \omega^2}{(\omega^2 + \omega_Y^2) * (\omega^2 + \omega_c^2) * (\omega^2 + \omega_s^2)} \qquad [\text{m}^2 / \text{rad/m})] \qquad (16)$$

where $S(\omega)$ represents the PSD of the different irregularities, ω represents the spatial wavenumber $2\pi/\lambda$ in [rad/m], ω_c, ω_Y, ω_s represent critical numbers in [rad/m], A_v represents the constant factor for the vertical profile (used also for cross level) in [m*rad], A_a represents the constant factor for the lateral alignment in [m*rad] and A_g represents the constant factor for the gauge in [m*rad].

Track class	ω_c [rad/m]	ω_Y [rad/m]	ω_s [rad/m]	A_a [10^{-7} m*rad]	A_v [10^{-7} m*rad]	A_g [10^{-7} m*rad]
Low irregularities	0.8246	0.0206	0.4380	2.119	4.032	0.532
High irregularities	0.8246	0.0206	0.4380	6.125	10.80	1.032

Table 14: Parameter values for the German PSD standard per (Lei 2017)

The FRA standard is based on a large number of field measurements fitted to a function expressed by cutoff frequencies and a roughness constant (Lei 2017). The FRA classifies the track irregularities in nine classes. Class one to six are dedicated to regular trains. Classes seven to nine are being developed for high speed services. The PSD model is characterized by a one-sided PSD function. The formulas and parameter values to determine the FRA curves are:

PSD vertical irregularities

$$S_v(\omega) = \frac{kA_v * \omega_c^2}{(\omega^2 + \omega_c^2) * \omega^2} \qquad [cm^2 / rad/m)] \qquad (17)$$

PSD lateral irregularities

$$S_a(\omega) = \frac{kA_a * \omega_c^2}{(\omega^2 + \omega_c^2) * \omega^2} \qquad [cm^2 / rad/m)] \qquad (18)$$

PSD cross level

$$S_{cl}(\omega) = \frac{4kA_v * \omega_c^2}{(\omega^2 + \omega_c^2) * (\omega^2 + \omega_s^2)} \qquad [cm^2 / rad/m)] \qquad (19)$$

PSD gauge

$$S_g(\omega) = \frac{4kA_v * \omega_c^2}{(\omega^2 + \omega_c^2) * (\omega^2 + \omega_s^2)} \qquad [cm^2 / rad/m)] \qquad (20)$$

where $S(\omega)$ represents the PSD of the different track irregularities, ω represents the spatial angular wavenumber in [rad/m], ω_c and ω_s represent critical wavenumbers in [rad/m], k represents a determined variable with value 0.25, A_v represents a roughness coefficient for the vertical profile (used also for cross level and gauge) in [cm^2 rad/m], A_a represents a roughness coefficient for the lateral alignment in [cm^2 rad/m]. Table 15 shows the values of the parameters per class and speed (Lei 2017; Berawi et al. 2010):

Line Class	Max. Line speed		A_v	A_a	ω_c	ω_s
	Freight [km/h]	Passenger [km/h]	[cm^2 rad/m]	[cm^2 rad/m]	[rad/m]	[rad/m]
1	16	24	1.2107	3.3634		0.6046
2	40	48	1.0181	1.2107		0.9308
3	64	97	0.6816	0.4128	0.8245	0.852
4	97	129	0.5376	0.3027		1.1312
5	129	145	0.2095	0.0762		0.8209
6	177		0.0339	0.0339		0.438

Table 15: **Parameter values for FRA PSD standard per (Lei 2017)**

Figure 19 shows a comparison of both standards. In general, the lower-class FRA PSDs display higher values per wavelength than the German PSD. This is as expected since as seen in Table 15 lower classes are dedicated to trains operating at lower speeds. These classes might not be suitable for LRT systems since they operate with an average speed of 30 km/h and a maximum of 80km/h.

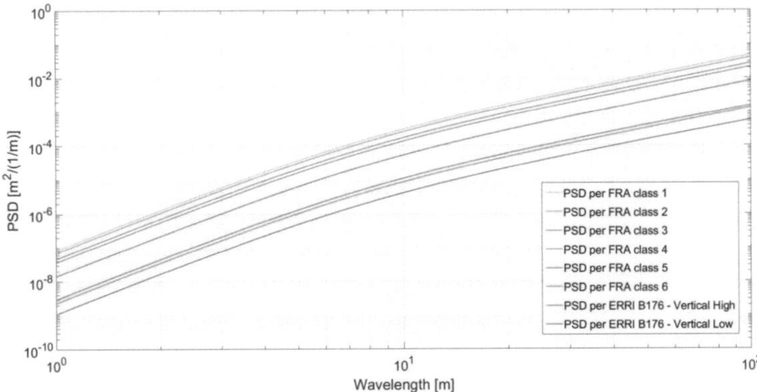

Figure 19: **PSD standard curves for vertical irregularities per ERRI B176 and FRA (own work)**

A further use of the PSD standard is the generation of synthetic signals that contain the frequency characteristics of a base signal through the inverse Fourier transformation. This aspect is explored further in section 3.4.1.

2.5.3.2 Time-Frequency Analysis

One distinctive aspect of the PSD is the loss of spatial information. Time-Frequency analysis (Spatial-Frequency) methods such as Short-Time Fourier Transform (STFT) and Wavelet analysis are used to perform frequency analysis while at the same time allowing wavelengths and amplitudes to be located along the track. Hence, time-frequency analysis provides a good overview of the track quality for discrete failures rather than an overall quality. As pointed out in (Ilvedson 1998), time-frequency analysis can help estimate the impulse response of a system, being able to distinguish the signal information that is due to system dynamics from that due to noise or disturbances.

In the case of the STFT, it can be seen as the inner product of a signal with a fixed window translating into both time and frequency (Jiang et al. 2012). In that sense, the STFT transforms short segments of spatial domain signals into frequency domain through sliding narrow windows across the signal producing larger frequency spectra while keeping the space, location, information of the harmonic details of the signal (Luber 2011). Since the STFT is based on the FFT of each segment is taken, frequency spreading is introduced into the results due the discontinuities at the ends of each segment (Luber 2011) (Ilvedson 1998). According to (Luber 2011) this is stronger when the window width is shorter that the period of the signal. To diminish the effect, a Hanning window is used in (Luber 2011; Ilvedson 1998).

$$W_{Hanning}[n] = 0.5 \left[1 - \cos\left(\frac{2\pi n}{N-1}\right)\right]$$
(21)

where $n = 0, 1, ..., N - 1$

The main disadvantage of the STFT is the resolution discrepancy between the time (location) and frequency information (Luber 2011) being introduced by the width of the window used. An increase in frequency resolution is obtained through a narrower window, while the opposite is true, a wider width of the window, increases the time resolution. Likewise, the width of the window influences the ability to determine the resolution of low and high frequencies. A sufficiently wide time window is required to appropriately characterize low-frequency components. Conversely, a narrow time window is appropriate to characterize high-frequency components (Di Scalea and McNamara 2004). Hence a constant width window is not able to provide an appropriate resolution for both high and low frequencies (i.e. wavenumbers) (Di Scalea and McNamara 2004).

The other time-frequency method is the Wavelet transformation (WT) which decomposes signals into dilated (scaled) and translated wavelets (Mallat 1999). A wavelet is a wave-like oscillation of finite duration and zero mean that functions as window function in the time and frequency domain (Di Scalea and McNamara 2004). The analysis procedure is to adopt a wavelet prototype function, called an analyzing wavelet or mother wavelet, which is translated across the signal and appropriately scaled to provide the required resolution. The prototype provides a high resolution in time and frequency (Di Scalea and McNamara 2004) deriving from a flexible window that is broader in time for observing low frequencies and shorter in time for observing high frequencies (Di Scalea and McNamara 2004). The wavelet transform is defined by the following expression:

$$Wf(u,s) = \int_{-\infty}^{\infty} f(t)\frac{1}{\sqrt{s}}\Psi^*\left(\frac{t-u}{s}\right)dt$$
(22)

where $\Psi^*(t)$ denotes the complex conjugate of the mother function $\Psi(t)$ defined as:

$$\Psi_{u,s}(t) = \frac{1}{\sqrt{s}} \Psi\left(\frac{t-u}{s}\right) \tag{23}$$

Where u shifts the wavelet in time (space) and s controls the wavelet frequency band-width (Di Scalea and McNamara 2004).

As pointed out in (Zhiping et al. 2010) the wavelet analysis can be used to accurately evaluate the quality of construction of a track and to guide railway track maintenance. Additionally, (Luber 2011) points out that WT is able to detect special discontinuities such as discrete failures on the track or determine the pattern of geometrical failures that can be later detected or analyzed.

Chapter 2 has laid the theory and methods for the determination of track geometry, leading to the determination of track quality. In the following chapters track geometry will be evaluated through statistical and frequency domain methods on available track data. The evaluation will be combined with simulated vehicle responses to determine a limit of track intervention / renewal for light rail systems.

3 Track Data

The methods presented to determine TGQ and track condition are based on the analysis of track irregularities. This chapter describes how the required data is gathered and treated. The track data was measured with a track recording vehicle (HuDe) for 11 inspection periods (one inspection period was not used in the analysis due to extreme data corruption). The inspection was carried out every six months between May 2013 and March 2018. The data were provided by SSB in excel files for two concrete-sleeper-ballasted-tracks in their inbound and outbound direction for a total of 9.3 kilometers in each direction (see Table 16).

Track No.	Directions	Length [km] each direction	No. of usable inspection periods
330	inbound/outbound	2.8	10
400	inbound/outbound	6.5	10

Table 16:   **Track data provided by SSB**

The 11 inspection periods were labeled day 0 (May 2013) to day 1756 (March 2018) depending on the inspection date as seen in Appendix IV: Track Data Provided and Nomenclature. However, according to the segmentation method (see section 4.3), mostly sections for which maintenance and renewal history were known, and that are free from crossings or turnouts, were used in the analysis. This resulted in about 7.0 kilometers used from which 500 meters were for track 330 on a straight and flat segment. The rest was on track 400 which contained long radii curves and slopes higher than 2.0%. Track 330 was mostly used to develop the methodology in this work while track 400 was used to prove it and develop the limits of intervention. The latter was the case since track 400 displayed in general higher track geometry irregularities than track 330.

Track 330 was built in 1986 with wooden sleepers. After 28 years, in August 2014, it was renewed. The renewed track section is 850 meters from which 590 meters do not contain turnouts or lay within stations. The renewal consisted of the replacement of the track's superstructure and substructure. The wooden sleepers were replaced with concrete sleepers due to deterioration. The slopes of the track are under 2% along the entire section and no vertical or horizontal curves are present, except for subtle changes in slope. Tamping to correct alignment was performed approximately one year and three months after renewal. No more information about maintenance was provided. In 2013, at the beginning of the data for the study, only two LRT lines operated on track 330 (i.e. U3 and U8). In 2016, a third line was introduced (i.e. U12). The track experiences about 6.8 million tons per year.

The history of the track is known in terms of renewal and measurements (inspections periods every six months) which helped in the development of this research.

Track 400 is built mainly on a slope between 2.0 and 6.5 %. The section used in this thesis contains tangents and curves with minimum radius of 96 m which should not affect the analysis of the track performed for D1 wavelengths. Only one light rail line operates on the track (i.e. U1). The track experiences about 2.6 million tons per year.

No more information was provided for this track, although through the analysis in this thesis, it appears that the track was improved / renewed around August or September 2013.

3.1 Track recording vehicle (TRV)

Track recoding vehicles (TRVs) are used to measure, process and store track data. In Europe, track recording vehicles need to comply with EN 13848-2, which specifies the minimum requirements for measuring principles and systems to produce comparable results regardless of the vehicle used (The British Standards Institution 2006). For signals to be useful, the standard specifies that signals should be repeatable under similar conditions (e.g. speed, measuring method, etc.), the signal should be reproducible under different conditions, they should be comparable to signals produced by other vehicles under the same conditions and should provide a precise location of parameters (The British Standards Institution 2006). TRVs take their measurements based on two main principles and parameters required, namely chord offset and inertial as shown in the sections ahead.

3.1.1 Chord offset / versine method

For lateral and vertical alignments, the chord offset method (chord system) is used. A chord system consists of a length (L_{chord}) carrying three devices that sense positions at three points on the rail. The versine or sagitta (s) is evaluated based on subtraction form the data obtained (Lewis 2011) as seen in Figure 20.

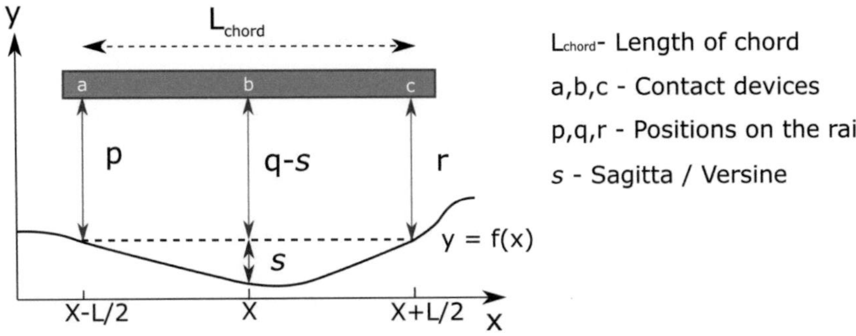

Figure 20:   **Chord measuring system – calculation of versine/sagitta per (Lewis 2011)**

Some systems need to compensate the car body bending with a virtual chord through a system of three laser-cameras located each on one of the measuring axes.

As explained in (Andrea Haigermoser 2013) chord systems have advantages and disadvantages. Chord based systems do not require a minimum speed to perform measurements. In addition, defects can easily be found, and the measuring principle is relatively simple. However, optical devices are required for non-contact sensors (devices); introducing the risk of failure (e.g. dirt on the lenses) and bad calibration. Moreover, as discussed in section 2.4.1, signals measured with chord systems need to be decolored to eliminate distortions introduced by the chord characteristics. For example, the HuDe system used in Stuttgart (see section 3.2), although asymmetrical, introduces distortions to measured amplitudes, which are either decreased or amplified as seen in the transfer function shown in Figure 21.

As it can be seen in the figure, 1.3-meter wavelengths are decreased about 0.2 times their real amplitude. The opposite occurs to amplitudes of 2-meter wavelengths which are increased about twice their real size. In addition, the HuDe system displays almost no ability to provide information above 50-meter wavelengths.

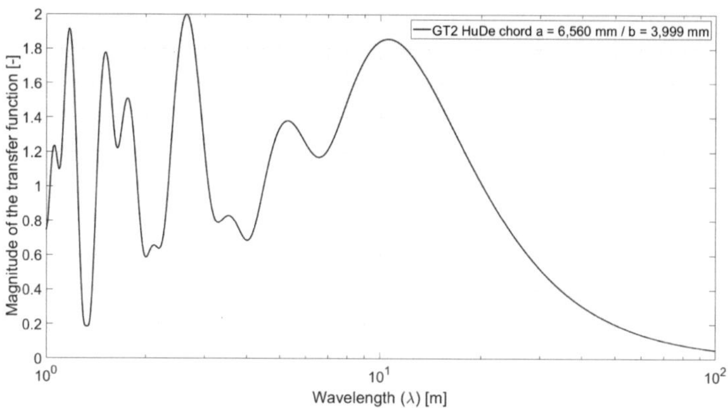

Figure 21: **Transfer function of the GT2 HuDe chord-system (own work)**

3.1.2 Inertial method

The inertial method consists of accelerometer and/or gyroscope sensors capable of identifying the displacement of a mass under inertial forces (Lewis 2011). The irregularities are obtained through the double integration of acceleration measurements (Lewis 2011; Chudzikiewicz et al. 2017) and the single integration of the rotational velocity measured by the gyroscope measurements (Benz 2016).

However, measuring systems cannot be used directly on railway systems due to a wide dynamic range of the signals. Carrying effective measurements inertial devices requires that they are placed on damped surfaces limiting the dynamic ranges (Lewis 2011).

(Andrea Haigermoser 2013) further indicates that an inertial method-based system using optical devices provides an absolute reference to the rail position in the vertical and horizontal directions. By means of the reference, the variation of the position of each rail in both directions can be calculated. However, the reference is not quite absolute and can only evaluate changes in position approximately between 3 and 200-meter wavelengths (Andrea Haigermoser 2013). In addition, sensors are susceptible to dust, grease, driving rain and snow (Andrea Haigermoser 2013; Lewis 2011).

An advantage of inertial systems includes the possibility to measure at high speeds. In addition, an inertial system is compact and fits on a car body or even a bogie frame. The main drawback is the minimum speed the train needs to move to measure the irregularities; which is between 10 and 30 km/h depending on the system (Andrea

Haigermoser 2013). Also, isolated effects get distorted and care must be exercised that filtering for the wavelength bands chosen do not introduce distortions to the signals (Andrea Haigermoser 2013).

3.2 Stuttgart's Track Recording Vehicle

In Stuttgart, the TRV is a combination of chord and inertial systems that include optical cameras and lasers. The vehicle consists of a locomotive unit called the HuDe GT2 "Gleismesssystem" and a control car containing the system's software (Figure 22). To determine the location (stationing) of irregularities, HuDe uses a radio frequency identification (RFID) reader that scans RFID tags located along the track (HuDe Mess- & Anlagentechnik GmbH 2011). The system also measures the speed of the vehicle, being positive when the locomotive hauls the control wagon and negative when the locomotive pushes it.

Figure 22:   **Stuttgart's Track Recording Vehicle - GT2 HuDe (photos: David Camacho)**

The inertial system is attached to the frame of the GT2 locomotive. The unit contains three accelerometers and three gyroscopes (HuDe Mess- & Anlagentechnik GmbH 2011). The accelerometers determine the vertical (a_z), horizontal (a_y) and longitudinal (a_x) accelerations. The gyroscopes determine angular rotations: $Kreisel\_X$, $Kreisel\_Y$ and $Kreisel\_Z$ which correspond respectively to the movements known as roll, pitch and yaw shown in Figure 23. The position is calculated through the double integration of the accelerations. The rotation angles are calculated with the single integration of the angular rotation.

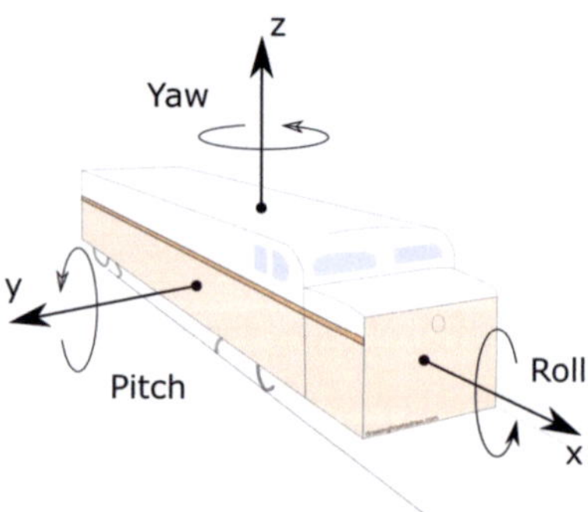

Figure 23:   **Angular measurements of an inertial measuring system (own work)**

The chord measuring system in the GT2 HuDe system consists of a virtual chord used for measuring horizontal irregularities. The total length of the chord (L_{chord}) is 10,259 mm divided in two segments a ($\overline{P1P2}$) and b ($\overline{P2P3}$) measuring 6,560 mm and 3,999 mm respectively as seen in Figure 24.

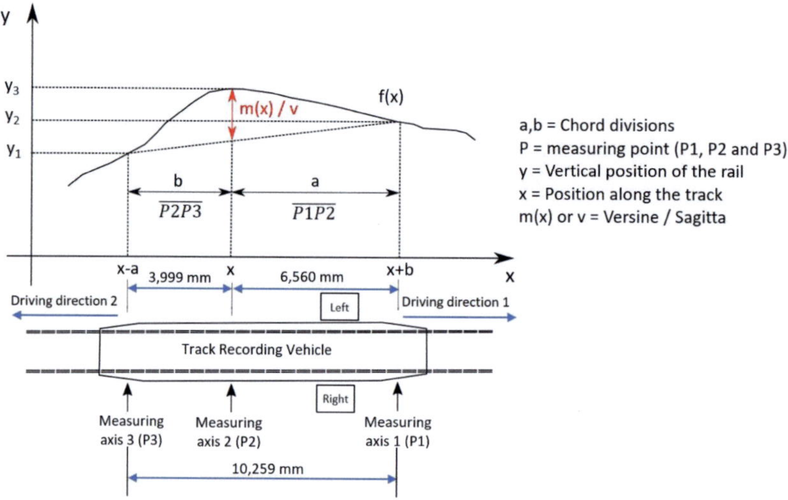

Figure 24:   **Chord measuring system configuration per (Andrea Haigermoser 2013)**

According to HuDe admin personnel, the measuring system used by SSB can apply filters to transform the parameters into frequency domain. Such measurements would be beneficial for the approach delineated in this research.

The HuDe system measures the track gauge, cross level/cant, twist/warp, lateral and longitudinal levels (for both rails), as well as layout or alignment parameters of the track (i.e. curvature). In addition, HuDe provides information about management aspects of the infrastructure such as stationing of the tracks for easier identification of irregularities or infrastructural elements (i.e. switches); this capability is accompanied by cameras to record the track condition. In the following paragraphs, the parameters measured by the HuDe are briefly discussed based on the information provided in the system's manual (HuDe Mess- & Anlagentechnik GmbH 2011) and (Benz 2016):

The **track gauge** (Ger. Spurweite) is measured in mm. The measurement is performed by the center measuring axis which determines the horizontal deflection of the rail heads via the left and right camera. The cameras take a two-dimensional image of the rails as a height profile. Based on the known relationships between pixels and distance data in the images, a deviation is determined from the horizontal reference position of the rail. The track width is calculated as the sum of the left and right horizontal deviations from the nominal gauge value. A two-dimensional image is recorded and compared with the correlation profiles of the rails deposited in the system.

The **longitudinal/vertical level** (Ger. Längshöhe) is given in mm for the centerline of the track. The vertical irregularity is determined based on the gyroscope parameter "$Kreisel\_Y$" (pitch). The zero to peak vertical values are determined measuring the inclination α through the gyroscope measure $Kreisel\_Y$ in radians for a measuring basis MB distance of 10 meters:

$$z = Tan\,(\alpha) * MB \tag{24}$$

This formula represents the relative vertical irregularity due to changes in slope on the track.

The **lateral/horizontal irregularity** in the HuDe system is measured for the left and right rail. The parameter measured is called **sagitta/versine** (Ger. Pfeilhöhe), measured in mm using the distance between the rail and the virtual chord system of length $\overline{P1P3}$ formed by the three camera-laser sensors. The sagitta provides the deviation of the track on the lateral direction as well as a characterization of curves present in the alignment. The measure has a range of ± 292 mm; corresponding to a curve radius of

45m. The distance between the virtual chord and the rail is determined through the middle point ($P2$) or sensor as shown in Figure 25.

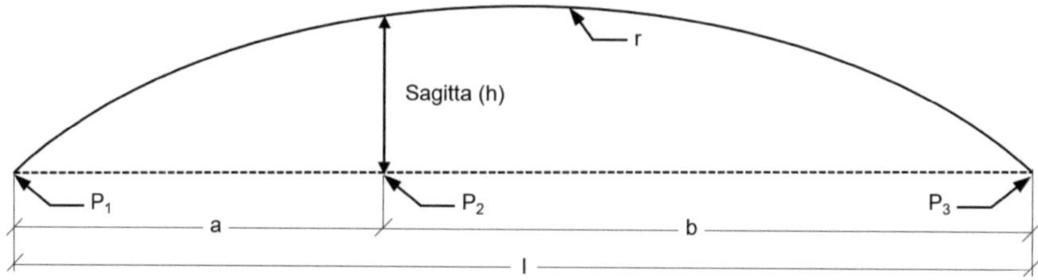

Figure 25:   **Chord system measuring principle per (HuDe Mess- & Anlagentechnik GmbH 2011)**

The equation to calculate the lateral irregularity is the following:

$$h = \frac{a * b}{2 * r} \tag{25}$$

where h represents the lateral irregularity (sagitta), $a = \overline{P1P2}$ and $b = \overline{P2P3}$ represent chord divisions, l represents the chord length and r represents the radius.

The ***cross level/cant*** (Ger. Gegenseitige Höhenlage) is measured by the gyroscope. The measuring range and basis is ± 1435 mm; being the gauge of the track in mm. The equation to calculate the cross level is the following

$$Tan\,(\theta) = \frac{CL}{1{,}435\;mm} \tag{26}$$

where θ represents the angle in radians measured and CL the cross level in mm

The ***twist*** (Ger. Verwindung) is calculated as the difference between two cross level values in a 10-meter distance:

$$Twist = CL_{staA} - Cl_{staB} \tag{27}$$

where CL_{staA} represents the cross-level value at point (station) A and CL_{staA} represents the value of cross-level at point (station) B.

3.3   Treatment of measured data

3.3.1   Spatial domain treatment

The data treatment prepares signals for their use in the frequency domain analysis. Figure 26 shows an example of the treatment process for the vertical irregularities (the diagrams for the other parameters can be found in Appendix XIII: Flow Charts for Formal Specifications).

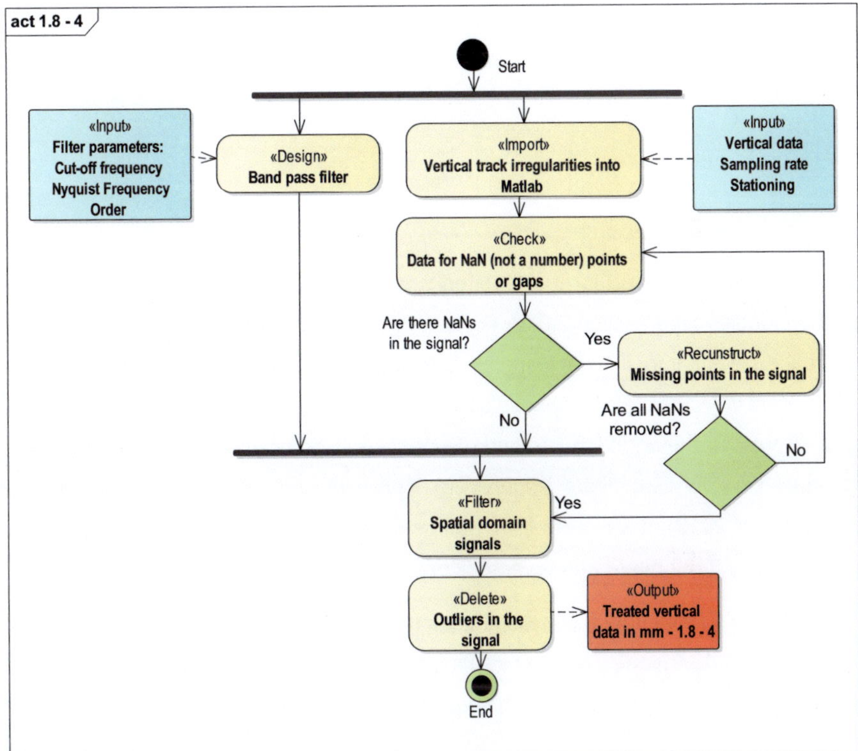

Figure 26:   **Data treatment process for vertical irregularities (own work)**

The first step is to check for corrupted or unmeasured data (i.e. NaN or not a number) to avoid errors in further stages of the analysis. To reconstruct missing data the fillgaps() MATLAB[©] algorithm called "Reconstructing Gaps with Localized Estimation" was used. The algorithm replaces NaN values with estimates extrapolated from forward and reverse autoregressive fits obtained from the remaining samples (Mathworks 2018b)

(Figure 27). Care must be exercised to avoid very long sections of missing data that introduce deterministic effects (e.g. interpolation).

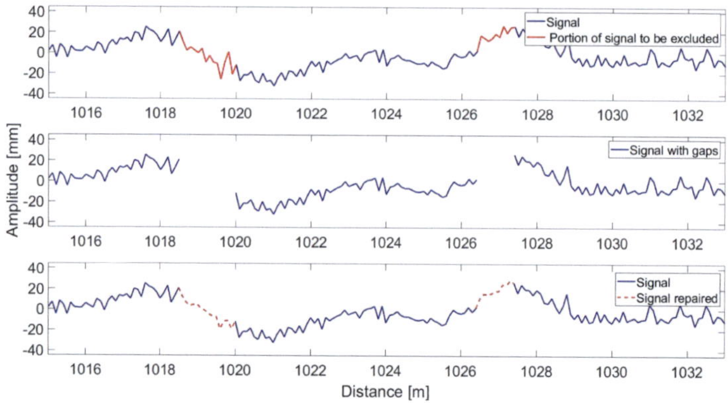

Figure 27:   **Reconstruction of signals (own work)**

Next the signal is filtered for the required wavelengths as required by the vehicle eigenfrequencies (see section 2.3.2). For the vertical irregularities, the filter is designed as a bandpass filter for D1 wavelengths. The filter design follows the recommendations established in (Andrea Haigermoser 2013) which mentions that filters should be based on a 4th order Butterworth forward and reverse filter in order to eliminate the phase shift introduced by forward filters (Figure 28). However, as recommended in (Andrea Haigermoser 2013), a 2nd order filter was used to account for a doubling order effect introduced by the forward and reverse filter.

Often signals display unrealistic amplitudes (i.e. outliers) that affect statistical analysis (e.g. standard deviation). The data treatment includes the elimination of outliers in the signal though a MATLAB$^©$ function *filloutliers()* which detects and replaces outliers in data more than three local scaled mean absolute deviation from the local median over a window length (e.g. 100 meters) (Mathworks 2018c).

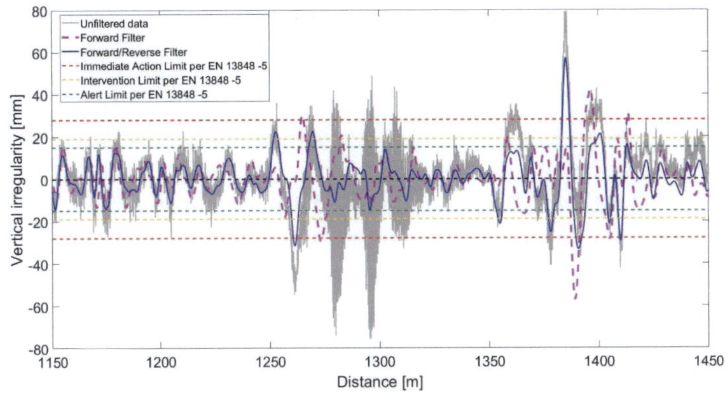

Figure 28:   **Data filtering in forward and reverse directions (own work)**

There are aspects of the process shared by all parameters (e.g. filtering, outlier removal, etc.) and specific aspects unique to each parameter as described in Table 17. The vertical and lateral parameters are filtered for D1 wavelengths corresponding to the wavelengths of interest. The lateral parameter requires a previous decoloring to reconstruct the signal distortion introduced by the measuring system. Both parameters are filtered with a forward and reverse Butterworth bandpass filter to prevent phase shifting (Andrea Haigermoser 2013). The cross level and the gauge do not required filtering for a specific bandwidth (Deutsches Institut für Normung e. V. 2008); however, as discussed in(Andrea Haigermoser 2013), applying the pwelch function for a non-filtered signal causes instability in the PSD as the number of segments increase[2]. (Andrea Haigermoser 2013) recommends that the signal is filtered with a high pass filter for 200-meter wavelengths. To control phase shift, forward and reverse filtering was also used.

---

[2] Higher number of segments is used to smooth the PSD curve.

Track irregularity	Type of filter	Wavelengths filtered [m]
Vertical	Bandpass 2nd order forward and reverse Butterworth filter	$3 - 25$
Horizontal	Bandpass 2nd order forward and reverse Butterworth filter	$3 - 25$
Cross Level	High-pass 2nd order forward and reverse Butterworth filter	200
Gauge	High-pass 2nd order forward and reverse Butterworth filter	200

Table 17: **Specifics of filters to treat track irregularity data (own work)**

The filtered vertical and horizontal signals for the worse quality observed, show that maximum and standard deviation limit values have in most cases been exceeded for the vertical irregularities, but not so for the lateral irregularities. Figure 29 shows the vertical irregularities for the worse day observed for track 330 (day 320) against the limit values as established in (The British Standards Institution 2010). As it can be observed, the irregularities either exceed or are close to the Immediate Intervention Limit IAL.

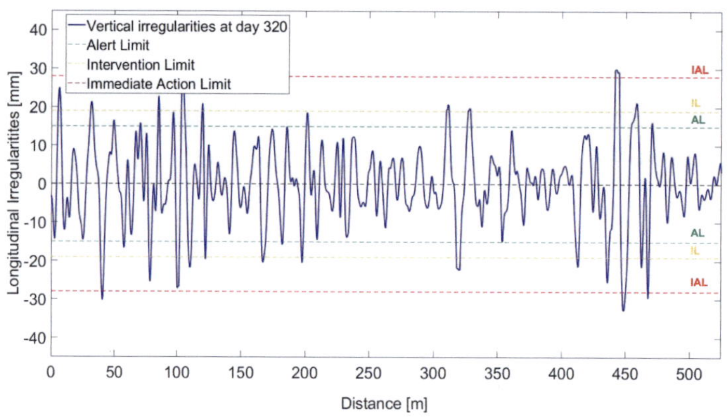

Figure 29: **Vertical irregularity of worse day observed against to limits per (The British Standards Institution 2010)**

Likewise, Figure 30 shows the same signal's standard deviation against the AL established in (The British Standards Institution 2010) for vehicles traveling at speed lower than 80 km/h. Here, the whole signal has surpassed the AL limit.

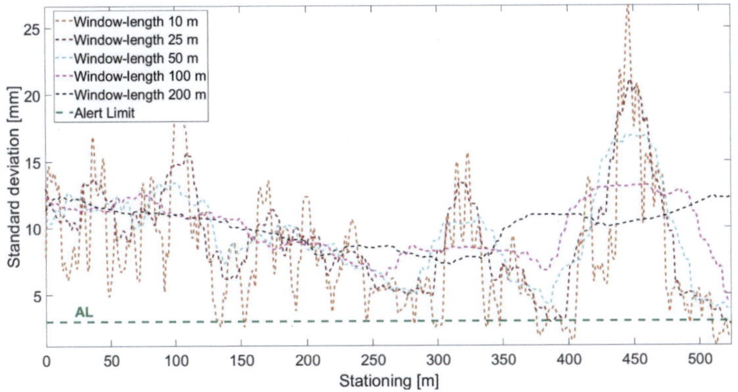

Figure 30:   Moving SD of worse vertical track geometry observed against limit per (The British Standards Institution 2010)

In the case of horizontal irregularities (see Figure 31), the signal is well below the Alert Limit for the worse day observed (day 320) for track 330.

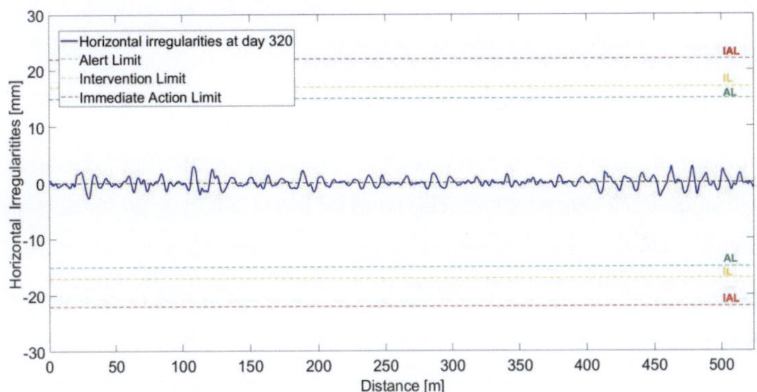

Figure 31:   Lateral irregularity of worse day observed against limits (own work)

A similar situation can be observed for the standard deviation limits as it can be seen in Figure 32.

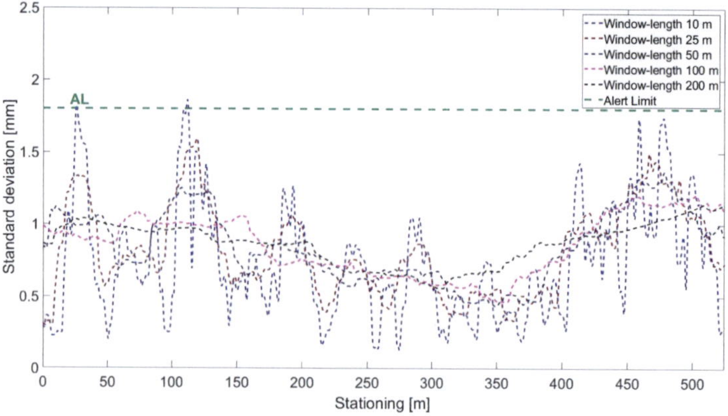

Figure 32:   **Moving SD of worse lateral track geometry observed against limit (own work)**

Gauge also displays a low SD in comparison to limits provided in (The British Standards Institution 2010). In the case of cross level, standard EN 13848-5 does not provide a limit value because the risk associated to cross level defect is related to twist (The British Standards Institution 2010).

3.3.2   Frequency domain analysis of data

Treated data in time domain can be used for the determination of frequency domain analysis through PSD which later can be used for the creation of synthetic irregularities used multibody simulations. The PSD calculation process for the vertical irregularities is depicted in Figure 33. The process for the other parameters is similar and can be reviewed in Appendix XIII: Flow Charts for Formal Specifications.

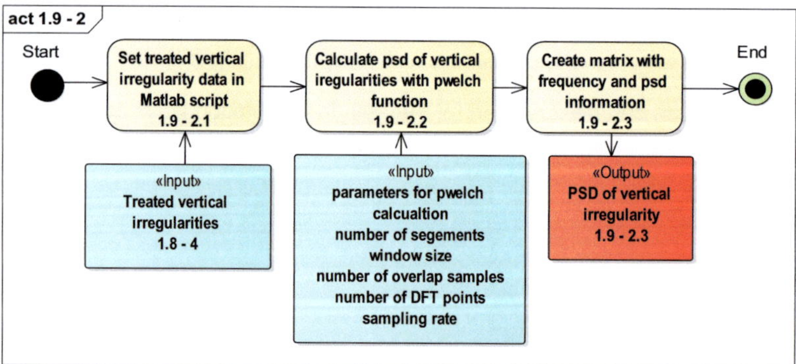

Figure 33:   **Process for the calculation of PSD for vertical irregularities (own work)**

As mentioned in section 2.5.3.1 the PSD was calculated using the pwelch function with 10 segments. Figure 34 shows the PSD of vertical irregularities for track 330 inbound for day 320 and 673, representing the worse and best track geometry observed.

As depicted in Figure 34, the PSD of the track, at its worse geometrical quality, surpasses the lowest quality of the FRA classes (i.e. class 1) for wavelengths between 3.4 m and 18 m. The best quality observed, after a track renewal, still displays a similar PSD curve but with lower values. Interesting is that for the wavelengths below 4.7 meters, the track still surpasses class 1. For shorter wavelengths, the track displays a worse geometrical quality. For wavelengths longer than 4.7 meters, the improvement was substantial, falling below class 4 and 5 for different wavelengths.

As also seen in Figure 34, the deterioration of the track in the vertical direction after renewal has not changed significantly in a three-year period. Furthermore, it can be observed that there is still some time until the worse observed quality is reached.

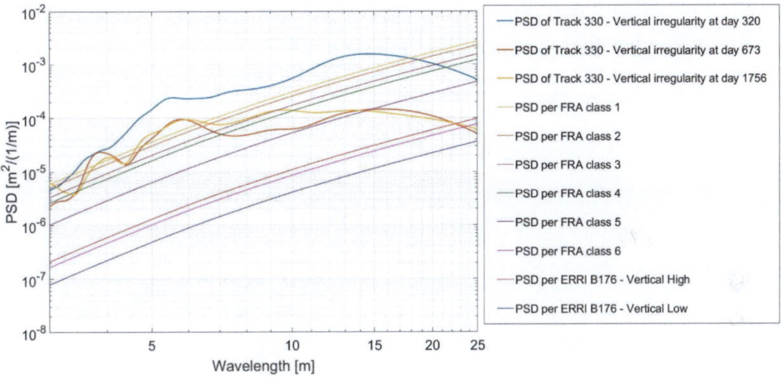

Figure 34: **PSD of vertical irregularities for the worse, best and latest track geometry quality (own work)**

A similar situation can be observed for the PSD of horizontal irregularities, but they do not exceed the FRA 1 class. They rather lay between class 4 and 5 (see Figure 35). An interesting aspect is that horizontal irregularities display a worse track geometry at day 1756 than at day 320. However, the track irregularity is still below class 4.

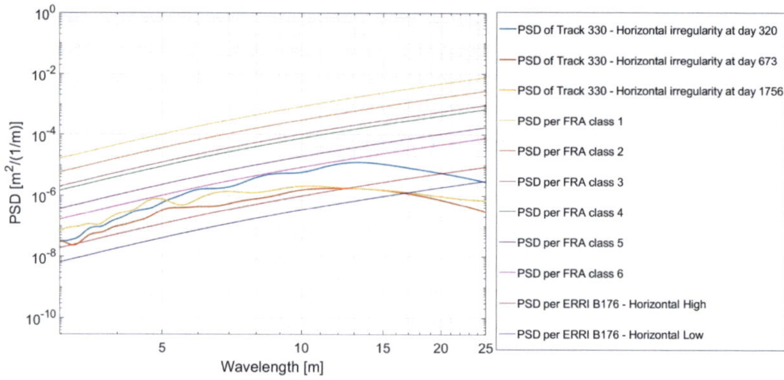

Figure 35: **PSD of horizontal irregularities for the worse, best and latest track geometry quality (own work)**

PSDs for the gauge and cross level present a special situation since these parameters do not require to be filtered. However, when the pwelch function is applied, the PSD curve deviates upward at about 30 m as shown in Figure 36. To avoid this situation, a 200-meter high pass filter is applied as recommended in (Andrea Haigermoser 2013).

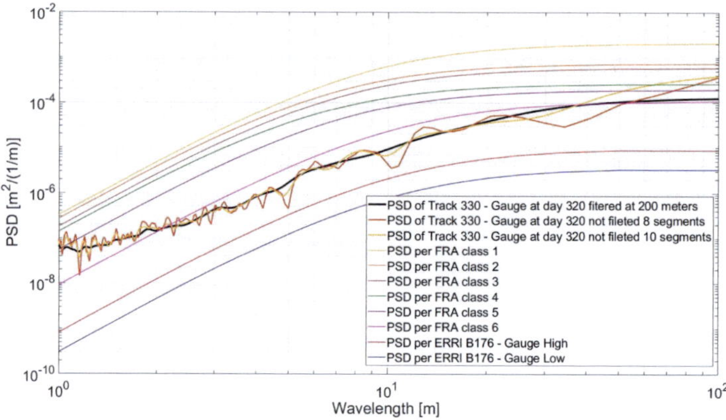

Figure 36: **PSD of gauge depicting deviation of curves due to lack of filtering (own work)**

As it can also be seen in Figure 36, the gauge PSD is close to FRA class 4, which represent a good geometrical quality for a slow speed system (i.e. LRT). The same situation can be observed for the cross level.

Once the gauge and cross level are filtered, they can be plotted in a PSD diagram without showing upward deviations. Figure 37 depicts a PSD plot of gauge, which shows that its worst quality is present for wavelengths below three meters. However, in general does not display a large problem being close to class 5.

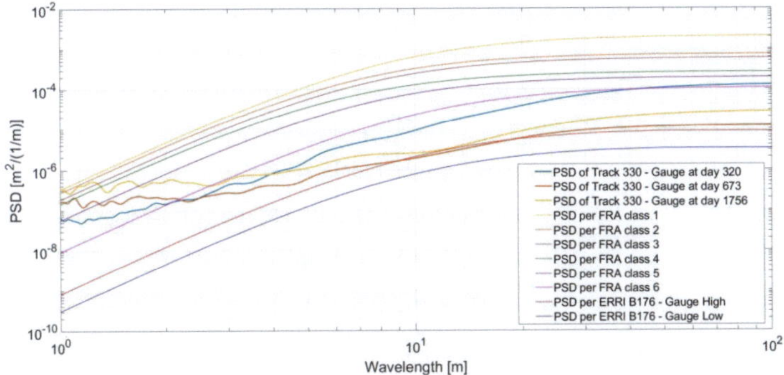

Figure 37: **PSD of gauge irregularities for the worse, best and latest track geometry quality (own work)**

Likewise, the cross level in Figure 38 shows higher irregularities below three meters, while keeping a very good quality for longer wavelengths.

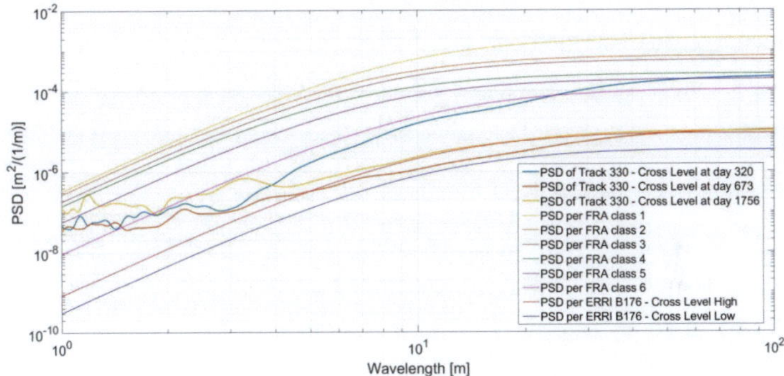

Figure 38: **PSD of cross level irregularities for the worse, best and latest track geometry quality (own work)**

3.4 Use of treated data

The measured data can be used for the creation of synthetic signals representing track irregularities, which can in turn be used in numerical simulations as the exciting force of the random vibration of the vehicle-track system (Lei 2017). Synthetic signal generation becomes more important in the absence of measured track irregularities, especially for tracks in bad conditions. In addition, being able to generate signals becomes important in the analysis of comfort level since the analysis requires a minimum of 300 seconds of running time (The British Standards Institution 2009) which as discussed later in section 5.1.2, it is not easily achieved in LRT infrastructure due to the short segments between stops.

3.4.1   Synthetic signals

Track irregularities are considered random processes based on time series of zero expectance and constant variance that describe stationary ergodic process in space (Podwórna 2015; Lei 2017). The random process represents the superposition of a series of harmonics with different wavelength, amplitude and phase (Lei 2017) which can be synthetically modeled. In (Dumitriu 2014) it is discussed that random track irregularities can be characterized statistically through spectral properties of track geometry determined from PSD functions.

The trigonometric Fourier series is a common method to generate signals that are used in numerical simulations involving the investigation of random vibration of the vehicle-track system (Dumitriu 2016). (Lei 2017) identifies four trigonometric methods used in numerous studies (Lestoille et al. 2015; Simpack News; Perrin et al. 2013; Podwórna 2015; Salcher 2015).The methods are characterized by the PSD of widely accepted standard PSD curves (see section 2.5.3.1) to produce spatial domain signals (track irregularities).

Random samples $r(x)$ from PSD functions are determine through the trigonometric series which is equivalent to an inverse Fourier transform (Dassault Systems Simulia Corp. 2017). However, since PSD functions contain no phase information, it needs to be introduced randomly assigning values in the range $(0\ 2\pi]$.

One trigonometric series commonly used is the following (Lei 2017; Podwórna 2015):

$$r(x) = \sum_{k=1}^{N} a_k \cos(\omega_k x + \phi_k) \tag{28}$$

where

a_k Gaussian random variable with mean 0 and standard deviation σ_k, independent of each other for $k = 1, 2, ..., N$

σ_k^2 Variance $2S_x(\omega_k)\Delta\omega^3$ for $k = 1, 2, ..., N$

ϕ_k Random phase angle uniformly distributed in the range $(0\ 2\pi]$ and independent of a_k and of each other for $k = 1, 2, ..., N$

ω_k Discrete angular frequency $(\omega_{min} + (n - 0.5))\,\Delta\omega$

$\Delta\omega$ Frequency increment $(\omega_{max} + \omega_{min})\,/N$

N Total number of frequency increments (harmonics) in $[\omega_{max} + \omega_{min}]$

ω_{min} Minimum angular frequency (wavenumber) $2\pi/\lambda_{max}$ where λ_{max} represents the upper limit of wavelength band analyzed

ω_{max} Maximum angular frequency (wavenumber) $2\pi/\lambda_{min}$ where λ_{min} represents the lower limit of the wavelength band analyzed

x Length of signal in increments of $1/N_s$ where N_s is the number of samples

Substituting a_k with σ_k the Equation 28 becomes:

$$r(x) = \sum_{k=1}^{N} \sqrt{2\,S_x(\omega_k)\Delta\omega}\,\cos(\omega_k x + \phi_k) \tag{29}$$

Other methods use either a sine function or combination of cosine and sine functions to generate signals (Podwórna 2015; Lei 2017; Berawi 2003).

As explained in (Podwórna 2015), to generate reliable and low calculation cost signals, the value of N (i.e. harmonics) should be large enough and the number of samples N_s should be kept to a minimum. (Podwórna 2015) presents a numerical method to determine appropriate values. The method is based on the calculation of the standard deviation $SD[\bar{r}]$ and of the expectance (average) $E[\bar{r}]$ of the maximum value of a synthetic signal $\bar{r}$ for a length x of 100-meters. From the analysis, recommended values are $N = 2000$ and $N_s = 300$; however, acceptable values are $N = 1500$, $N_s = 100$.

---

[3] In literature the variance is provided as $4S_x(\omega_k)\Delta\omega$, however, through simulation of several signals, it was found that the value of 4 resulted in higher irregularities than expected.

Figure 39 shows the result of the analysis, where the expectance of the maximum value and standard deviation of the signal stabilize at $N = 1500$ for $N_s = 200$.

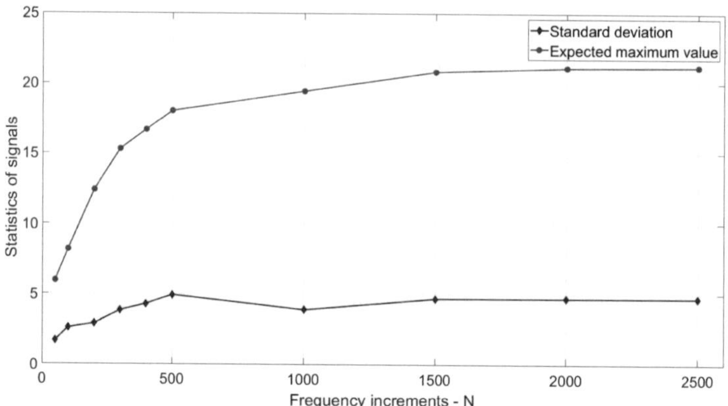

Figure 39:   **Statistics of signals for N_s = 200 (own work)**

Similarly, in Appendix V: Synthetic Signal Generation, $N_s = 100$ and $N_s = 300$ show that stability is achieved at $N = 1500$. However, for $N_s = 50$ stability is not reached for the maximum value of frequency increments analyzed (i.e. $N = 2500$), which provides an indication that N_s should be above 100.

3.4.2   Synthetic data in simulations

A tool to generate signals is built in the MBS software used for simulations in this study (i.e. SIMPACK©). SIMPACK© generates signals called excitations using the method called stochastic from PSD. As mentioned in 3.4.1, this excitation method takes a PSD as input and creates distance domain signal, which can be applied directly to the track. In SIMPACK© the sample needs to be sampled at given discrete frequencies n_F in a given frequency band F_{min} and F_{max}. For each sample frequency the amplitude is determined from the PSD $S(\Omega)$ as follows:

$$A(\Omega) = \sqrt{\frac{S(\Omega)}{2\pi\,\Delta\Omega} * \Delta\Omega} \tag{30}$$

where

Ω is the angular frequency $2\pi F$ being F the frequency (wavenumber) in 1/m

$\Delta\Omega$ is the angular frequency increment $2\pi\Delta F$ being ΔF the frequency increment

From the resulting amplitudes and phases, harmonic functions are created and summed up (i.e. trigonometric Fourier method) generating distance domain signals.

The method used in SIMPACK© produces equivalent results as the trigonometric Fourier method discussed above (Figure 40). As shown in Table 18, the statistical values that resulted from both synthetic signals generated are similar.

	Maximum expected value	Standard deviation
Synthetic signal (Trigonometric function)	22.9	8.0
Synthetic signal (SIMPACK©)	20.4	8.3

Table 18:   **Statistics of synthetic signals generated with two different methods**

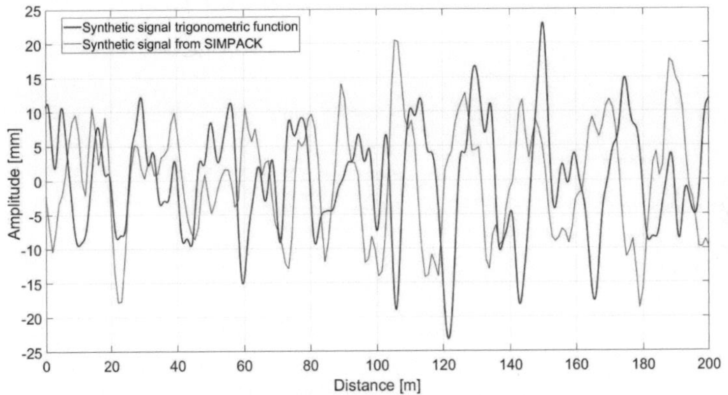

Figure 40:   **Signals generated through a Fourier trigonometric function and SIMPACK© (own work)**

Given that the signals generated in SIMPACK© are similar to the ones generated through the trigonometric methods, SIMPACK© is deemed acceptable for the generation of signals required for analysis. This study uses a value of $N = 2500$ and $N_s = 200$ to produce the signals.

3.4.3   Track Geometry Index (TGI) for LRT systems

As mentioned in section 2.5.1.2, the TGI is a practical and reliable method for the determination of TGQ. It also provides a great tool for infrastructure managers to visualize the condition of a track and its deterioration process. Depending on the amount of data, it also offers a relatively reliable prognosis of the deterioration for a short-term period (of three to five years as seen in section 5.5). Such prognosis might provide infrastructure managers with enough reaction time to plan maintenance and track renewal activities. For these reasons, the TGI method was used in this work for the analysis of the LRT tracks studied.

As indicated previously, the TGI formula reflects the influence of each parameter on an Indian Ride Index, in the case of the LRT under study the vertical parameter displays a larger influence than the lateral irregularity (Araji 2018; Álvarez 2018). This is due to the lower level of deterioration of the lateral irregularities compared to the vertical irregularities as seen in Figure 41 which shows the standard deviation of both parameters for tracks 330 and 400.

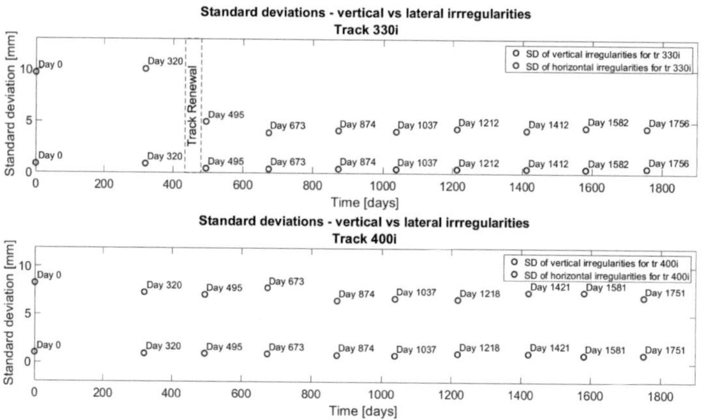

Figure 41:   **Vertical vs horizontal SD for tracks 330i and 400i (own work)**

As it can be seen, vertical irregularities have a higher SD than horizontal irregularities for both track and all days. In addition, vertical irregularities deteriorate over time faster than the lateral irregularities. This is more evident for the SD deviations before the renewal of track 330. Before renewal the SD of the vertical irregularities was in the range of 10, after renewal the SD was reduced to below 5 SDs. The lower level of deterioration of lateral irregularities can be also observed in the PSD plots in Figure 34 and Figure 35 where the lateral irregularity PSD remains very much lower than that of the vertical PSD (i.e. about 10 times lower).

Since the TGI for regular and LRT systems in terms of the influence of vertical and lateral irregularities, i.e. the influence of the parameters had no discrepancy, the formula of the TGI with no alterations was used in this study. However, since the systems are inherently different new SD values for a newly laid track and urgent maintenance were used in the calculation of the parameters' indices.

The values for a newly laid track were obtained from data corresponding to a renewed track (i.e. track 330). The urgent maintenance values were obtained from the standard deviation of synthetic data generated in the MBS that caused an uncomfortable ride (see section 5.3). Table 19 shows a comparison of standard deviations between the values from the regular TGI and the values used in this study.

Parameter	SD of regular TGI newly laid track [mm]	SD of LRT TGI newly laid track [mm]	SD of regular TGI urgent maintenance [mm]	SD of LRT TGI urgent maintenance [mm]
Uneven-	2.50	4.32	7.20	20.15
Twist	1.75	0.88	4.20	15.78
Gauge	1.00	0.98	3.60	4.39
Alignment	1.50	0.43	3.00	2.52

Table 19: **TGI values of regular trains and TGI values for LRT system under study**

4 Track Construction and Maintenance

This chapter provides a general idea of the construction of the ballasted track and its maintenance for LRT systems. The maintenance part is explored under the understanding that transit agencies differ in maintenance philosophy, policy and budgets (Daniels 2008). The chapter starts with a general review of the ballasted track construction and the way the structural quality of the track can be established. The chapter also explores existing maintenance philosophies and strategies as well as the lack of maintenance limits whose purpose should be to provide an idea of the track condition that asset / infrastructure managers could use as decision making tools or to develop a maintenance strategy.

4.1 Ballasted track construction

LRT systems present two different types of track formations, namely, embedded (Verband Deutscher Verkehrsunternehmen 2014) and separate/independent (Verband Deutscher Verkehrsunternehmen 2014). The embedded track formation is recognized as a covered track while the separate/independent track formations are considered open tracks (Verband Deutscher Verkehrsunternehmen 2014). The latter include ballasted, direct fixation ("ballastless"), and grass (turf) tracks (Esveld; Verband Deutscher Verkehrsunternehmen 2014; Parsons 2012). A mix of both types is recognized in (Verband Öffentlicher Verkehrsbetriebe 2008). This study focuses on LRT ballasted tracks, which in general are similar to ballasted tracks for conventional railways (e.g. freight) but present some important differences (Parsons 2012) that need to be taken into account (e.g. distance between sleepers, size of rail and sleepers).

In general, the function of a track is to guide the riding vehicle, however, in terms of structural support, the function of a track is to absorb and distribute static and dynamic stresses without permanently deforming the track for the maximum speed and loads allowed on a particular section (Verband Öffentlicher Verkehrsbetriebe 2008). As pointed out in (Indraratna et al. 2009), the ballast plays an important role in the track structure, namely, (i) distribution of load from sleepers, (ii) provision of dynamic load damping, (iii) provision of lateral resistance, and (iv) provision of free draining conditions. Tracks designed in a similar manner, would tend to provide similar carrying capacities and performance under the similar conditions. In this sense, when an LRT track is built structurally similar to a conventional railway track, it is expected to perform adequately for a longer period since for example a "LRT track structural loading is one-quarter to one-third of that imposed on freight railway tracks" (Parsons 2012).

The track structure can be represented by a continuous beam on springs which allows a deflection of the track in terms of the so-called track modulus C (Kerr 2002). Ballasted tracks provide the highest degree of flexibility among the available railway tracks (e.g. embedded), presenting a lower track modulus value in comparison with embedded tracks. This means that higher deflections under the same load can be observed as well as a more spread-out distribution of the pressure to the ballast and subgrade (Parsons 2012). Additionally, track modulus is known to lose vertical support as loads increase, which represents an issue in freight systems, but is unlikely to occur in transit systems unless weak soils are present (Parsons 2012).

A ballasted track consists of two main parts, the superstructure and the substructure[4]. The superstructure contains the main load-supporting elements that react and transfer the train load to the substructure (Li et al. 2016). The superstructure is required to withstand the vertical and lateral loads with small elastic and permanent deformation (Li et al. 2016). On the other hand, the substructure acts as the foundation of the track; supporting the superstructure and drainage elements (Li et al. 2016) and consists of ballast, sub-ballast and subgrade[5]. For this work, the superstructure consists of the rails, fastening system, elastomeric pads and sleepers while the substructure consists of the ballast, sub-ballast and subgrade as seen in Figure 42.

---

[4] Depending on the literature consulted, the superstructure and substructure are defined differently. For example, in [Kennedy (2011)] the superstructure consists of rails, fastening system, elastomeric pads, sleepers, ballast and sub-ballast while in [Hesse et al. (2014), Li et al. (2016)] the superstructure consists exclusively of rails, fastening system, elastomeric pads and sleepers. However, in [Tzanakakis (2013)] the superstructure includes the rails, fastening systems, rail pads, sleepers and ballast. The substructure presents similar definition issues. In some publications it is divided into formation layer and subgrade [Hesse et al. (2014)] while others consider the ballast and sub-ballast and consider the formation layer to be synonymous of subgrade [Li et al. (2016)].

[5] Subgrade also known as trackbed, roadbed track foundation and formation [Li et al. (2016)].

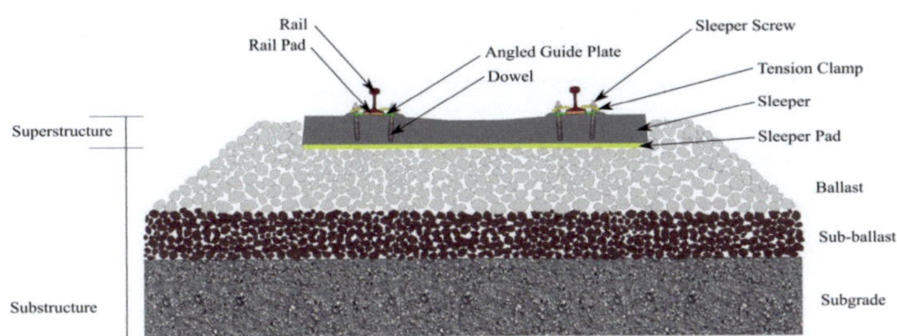

Figure 42: **Cross section of typical concrete sleeper ballasted track (own work)**

The structural carrying capacity and load distribution of the track greatly depends on the configuration of the individual components (e.g. types and separation of sleepers, grading of ballast stones, etc.). Likewise, each component contributes to the total stiffness of the track in the vertical direction measured with what is known the track modulus (Figure 43).

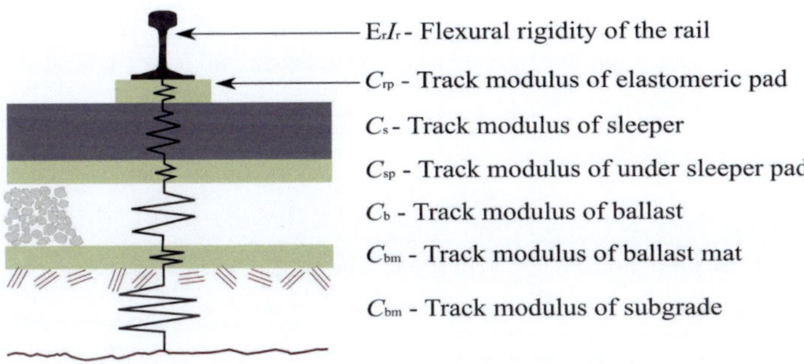

Figure 43: **Composition of total track modulus C per (Martin et al. 2016; Kerr 2002)**

The total track modulus describes then the structural condition of the track in the vertical direction (Martin et al. 2016). More formally speaking, the track modulus is the behavior of the stressed area under the sleeper in relation to the deflection of the ground (Martin et al. 2016). The track modulus depends on the quality of the track, the track's composition and construction and the influence of the weather conditions such as temperature (Martin et al. 2016), hence, the track modulus can be taken as an indicator of the structural quality of the track (Parsons 2012; Powrie and Louis 2016).

This research does not require the determination of the modulus of a track, and since construction practices in urban rail bounded transport systems and quality control in Germany is considered relatively high, the tracks used in simulations in this research were assumed to have a good structural quality. Table 20 shows the values of track modulus in relation to the structural quality per (Göbel and Lieberenz; Rapp 2017; Esveld).

Track Modulus [MN/m^3]	Track Structural Quality
50	Poor
100	Good
150	Very Good

Table 20:   **Track modulus of different track structural qualities per (Parsons 2012; Powrie and Louis 2016; Martin et al. 2016)**

4.2   Ballasted track of the system under study

Ballasted tracks are the most common track system in LRT systems when no street operation is required. This is sensible from the point of view of investment costs (Praticò and Giunta 2016), maintenance costs[6] and technical aspects (e.g. elasticity and damping properties are better than slab tracks) (Parsons 2012). This is in part due the lower weight and speeds of LRVs which translates into lower lateral and vertical loads imposed on the track (Parsons 2012), including for situations of relatively high track geometry irregularities as shown in section 5.2.

The light rail system under study consists of 127.55 km of tracks form which 68.68 km are ballasted tracks (Benz 2016). The track uses a 49E1 rail type while the sleepers are either wood, steel or concrete spaced 65 cm on center. Concrete sleepers are type B70 with a length of 2.4 meters and a width of 30 cm. Tracks in this study consist of 15-meter continuous weld rail (CWR) and concrete sleeper track panels.

Under the rail, an elastomeric pad is installed to decrease track stiffness (Powrie and Louis 2016), reduce impact energy and dampen dynamic frequencies (Kaewunruen and Remennikov 2008). The ballast layer is built with certified ballast in accordance to EN 13450. Certified ballast is commonly used in Germany in conventional railways; thus, it provides a carrying capacity and pressure distribution similar to conventional tracks (e.g. freight lines). As pointed out in (Benz 2016) the sub-ballast is built with a

---

[6] This statement does not apply to conventional lines for which the suitability of slab tracks in terms of LCC increases as speeds and traffic increase as shown in [Li et al. (2016), Praticò and Giunta (2016); Praticò and Giunta (2017)].

so-called "Frost Protection and Carrying Layer" (in Ger. KFT), which consists of a mixture of sand and gravel. Both, the ballast and the sub ballast are built with a minimum thickness of 25 cm. A typical section of a ballasted track for the system under study is shown in Figure 44.

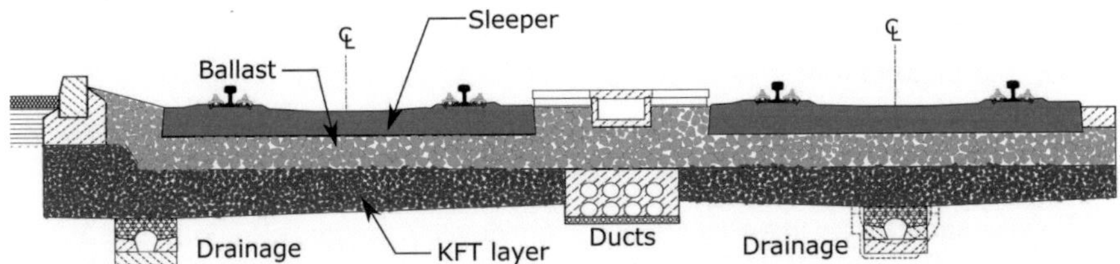

Figure 44: **Typical cross section of ballasted track for the system under study per (Benz 2016)**

Using certified ballast for regular railway systems generates a track substructure similar to a conventional railway system, which according to (Li et al. 2016) allows a sub-grade stress of up to 138 kN/m². Comparisons between resulting pressures form simulation at different speeds to allowable stresses will be performed in section 5.2. A comparison of the pressures against the allowable pressure at different depths, allows us to see whether pressures exerted are a factor to consider in the determination of maintenance limits.

4.3 Maintenance Strategies

The aim of maintenance is to keep a system within a level of functionality for the longest time possible by preventing it from rapid deterioration, which might require difficult and expensive corrective measures (Parsons 2012).

In general, there are four distinctive types of maintenance strategies, namely, run to failure (RTF), preventive (PM), corrective (CM) and predictive (PDM) (Tzanakakis 2013). As reported in (Daniels 2008), most transportation companies adhere to a preventive maintenance philosophy which might require a balance between prevention and correction measures due to budget constrain. Yet, the development of a maintenance strategy that relies on empirical planning, either by predetermination of maintenance cycles or interventions addressing symptoms of faults rather than their causes, might represent rapid recurrence of faults and constant and costly maintenance cycles

(Tzanakakis 2013). Additionally, an empirical and predetermined maintenance approach does not offer the possibility to optimize the track performance in terms of maintenance budgetary and resource constraints.

Furthermore, the development of a maintenance strategy accounting for track quality needs to consider the initial capital investment and renewal strategy (Veit 2007). In this sense, it is important to initiate with a good track quality that is sustained through maintenance regimes which at the same time slows down the deterioration rate and the track's degradation. Achieving this aim is possible when maintenance is adequately and timely carried out. Otherwise, tracks might develop irreversible shortening of their period of geometrical stability. When considering a maintenance regime, it is important to recognize that the initial track quality can never be achieved through maintenance. Nonetheless, it has been shown that a good quality deteriorates slower (with due maintenance diligence) than a poor quality or put in other words the current condition dictates the rate of deterioration (Veit 2007).

To be able to optimize resources, maintenance strategies, should consider systematic and intelligent ways to determine the appropriate type of maintenance as well as the frequency that maintenance should be performed (Tzanakakis 2013). In specific, it is necessary to create a condition and knowledge-based maintenance approach with certain predictive capabilities. Figure 45 shows a transition from a current maintenance approach a new approach seeking to increase the efficiency of the track.

In line with maintenance management principles, and the previous discussion, transport companies in Germany have developed an innovative maintenance management approach that establishes a framework for the conception, implementation and application of an innovative and economically sensitive, goal-oriented, maintenance management of light rail systems (Kochs and Marx 2009). Within this approach, maintenance management is defined as the collection of measures for the design, control and development of maintenance. The innovative maintenance scheme seeks to establish an efficient and goal-oriented maintenance management based on the use of data of the track at different points in time. The framework contemplates a combination of planned maintenance strategies as well as condition-based and emergency maintenance measures.

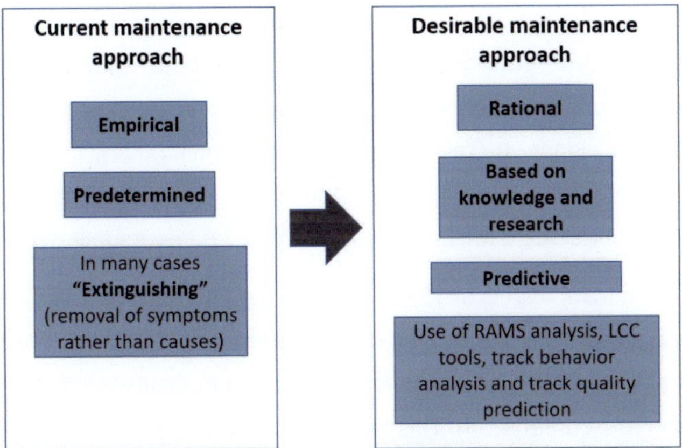

Figure 45: Transition from a current empirical approach to conduct maintenance to a rational-knowledge-based approach per (Tzanakakis 2013)

Moreover, the approach recognizes that intervention levels and prognosis of track failures (deterioration) are important aspects of a maintenance strategy. Such intervention levels should be used to assess the TGQ taking into account safety against derailment, the evaluation of the vehicle during acceptance tests and the durability of wheelsets and bogies (Kochs and Marx 2009). The document establishes requirements similar to regular railway services, namely, high initial track geometric quality (Veit 2007), timely replacement of old and expensive systems, condition-based maintenance, use of low maintenance and maintenance friendly systems, and standardization (Kochs and Marx 2009).

The strategy for an innovative maintenance management requires the use of an inventory of the infrastructure and its operation (e.g. including costs). The process relies on visual and metrological measurements or inspections as well as the quality assessment of maintenance works. In fact, it is through a metrological inspection using a TRV (see section 3.2) that the data used in this study is obtained.

To determine the frequency of maintenance, preestablished limits of intervention are required. These limits should allow a satisfactory ride quality in terms of passenger comfort for different speeds (e.g. average speed, and representative speeds), but should not allow the track to deteriorate to the point that safety is compromised (i.e. IAL), which is mostly achieved when preventive or condition-based maintenance are timely and accordingly performed.

To establish the limits that allow the design of maintenance strategies and planning, many aspects should be considered, among them the frequency of inspection (at least every 18 months for main lines running less than 80 km/h according to (DB Netz AG: Bautechnik, Leit-, Signal-, u. Telekommunikationstechnik 2004)) and record keeping that allows the reconstruction of the track's "history". In this research the two aspects are important to understand the track's deterioration and degradation process.

However, even if track data is available, to effectively manage and use it in a sensible manner, the infrastructure needs to be divided into segments having similar characteristics, age, track layout (e.g. mostly curves or mostly tangents) and track "history" (e.g. time of maintenance or renewal). This method is called segmentation process in which every segment is treated as a separate "organism" analyzed independently (Tzanakakis 2013).

4.4 Track segmentation

This study is based on the so-called track 330 and track 400 in accordance to the nomenclature provided by the local transportation company.

Track 330 was segmented according to its geometrical characteristics and type (e.g. conventional ballasted track construction with concrete sleepers) as well as its renewal information to be able to reconstruct its history (e.g. the TGI as seen in Figure 71).

If segmentation is not performed different track histories results in a distorted standard deviation as it can be seen in Figure 46. As it can be observed, the track under study shows a lower SD value after renewal under segmentation, while it shows higher standard deviations otherwise. As it will be seen later for the track geometry index, which depends on SD values, track segmentation is also important.

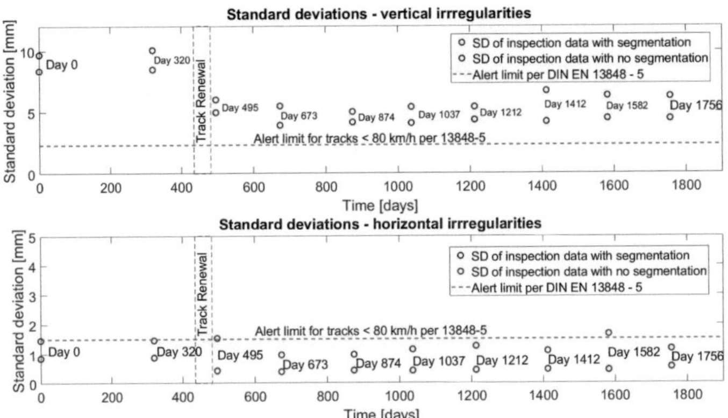

Figure 46: Effects of segmentation on a track with known history (own work)

Track 400 was not analyzed through segmentation because the history of the track was in general not known. In addition, it was necessary to find the worst track geometry possible (i.e. resulting from no segmentation as seen in Figure 46) to generate higher and synthetic track irregularities. This was done by increasing the magnitude of the PSD of the signals (see section 5.3). For the development of higher limits, track 400 was used since in general displays a worse TGQ (in terms of SD) than track 330 (see Figure 47), which might be in part due to the lack of track segmentation.

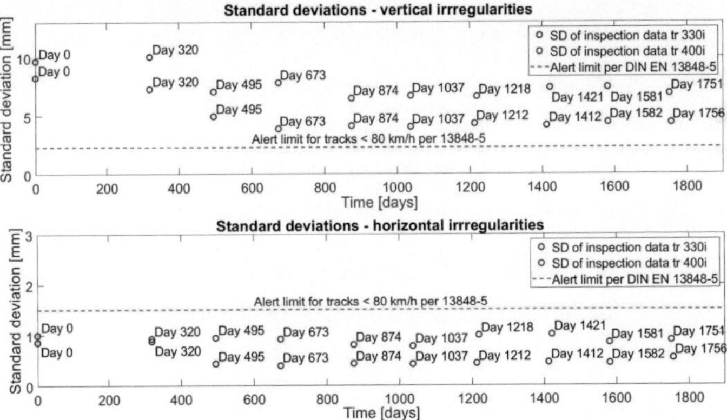

Figure 47: SD development for tr 330 and tr 400 depicting the difference in geometry quality (own work)

4.5 Vehicle performance as indicator of track maintenance needs

According to (Garg and Dukkipati 1984) the performance of railway vehicles is often defined in terms of safety and productivity. Safety is concern with derailment, which occurs for various reasons, including equipment failures, poor track quality, improper train handling, etc. In (The British Standards Institution 2016), the criteria to evaluate running safety are the sum of the maximum guiding forces ΣY_{max} which is used for assessing compliance in regards to safety against track shifting (The British Standards Institution 2016). The other common limit is the derailment ratio $(Y/Q_{a,max})$ of the leading wheel resulting from the climbing of the wheel flange onto the rail (The British Standards Institution 2016). Furthermore, the dynamic performance, as it is related to safety, is evaluated in terms of indices such as ride quality, vehicle stability, and vehicle curve-negotiation capability. An important input affecting the vehicles dynamics are the track geometry variations. As such, track irregularities are necessary for the evaluation of TGQ, vehicle performance, comfort and track lading (Garg and Dukkipati 1984).

Track productivity is related to operating and maintenance costs which are affected by operating speed as well as by track lading (Garg and Dukkipati 1984). Track lading is assessed through the quasi-static vertical wheel force $(Q_{v,qst})$, the maximum vertical load (Q_{max}) and the quasi-static guiding force $(Q_{a,qst})$, among others.

Although running safety is indeed important for railway systems in general, what has been observed for the LRT system under study is that track lading, dynamic coefficient and derailment coefficient do not to play initially a significant role in the determination of track maintenance limits since as discussed below and shown later in section 5.2, their values are significantly lower than the limits established in standards.

For example, simulations of the LRT system under study, have shown that at 80 km/h (maximum speed) the highest force experienced on the track at the worst track geometry condition observed is 70.52 kN[7] (see section 5.2). A calculation to check the pressure at the subgrade shows that the maximum force does not exert a pressure higher than the allowable at 50 and 60 cm below the sleeper (see Appendix X: Vertical Stress

---

[7] Maximum value observed was lower, the value reported here is based on a SD depending on the speed of the vehicle (in this case the maximum speed) and a statistically maximum safety factor see Appendix X: Vertical Stress Calculation

Calculation). Likewise, it is also noticeable that the maximum Y/Q ratio observed for the worst quality, is well below the derailment value of 0.8 (The British Standards Institution 2016) (see section 5.2).

However, the influence of these parameters might change with the increase of magnitude of track irregularities. This is carefully rechecked at later stages of the research.

Since initially safety and track damaging parameters are not close to established limits within the maximum operational speed and worst TGQ, it is necessary to evaluate another parameter to provide an indication of track deterioration and degradation for the LRT system under study. The evaluation can initially be performed through the ride quality, which it is interpreted as the capability of the vehicle suspension to maintain motion within the range of human comfort and prevent lading damage (Garg and Dukkipati 1984). The ride quality of a vehicle depends on displacement, acceleration, rate of change of acceleration (jerk) (Garg and Dukkipati 1984), and other factors not relevant for this research (e.g. noise). As pointed out in (Garg and Dukkipati 1984), two approaches are used to evaluate the ride quality of a vehicle: the fatigue time method and the ride index method. These methods determine the ride quality which is an evaluation of the acceleration of the vehicle or ride comfort, which implies that the vehicle is to be assessed according to the effect that mechanical vibrations have on occupants (Garg and Dukkipati 1984) This is represented by the time duration that an average person traveling in a vehicle starts having a distinct feeling of fatigue (Garg and Dukkipati 1984)

As mentioned in section 2.4.5, an index to evaluate the ride comfort is established in EN 12299. Contrary to the parameters for safety and track lading, the index shows a variation in its value that surpasses the comfort limit established in the standard at maximum speed and worse track geometry (see section 5.2). According to EN 12299, it is necessary to evaluate a 300 second of running time to establish the fatigue time of people due to their exposure to mechanical vibration. However, a time of fatigue might not be experienced in an LRT system since trains do not ride long periods of time to generate uncomfortable sensations. The index does not really attempt to establish how comfortable the passengers would feel within a riding vehicle, but rather provides an indication of poor run quality and track deterioration, which makes use of the same scale of passenger comfort, namely very comfortable, comfortable, medium comfort, uncomfortable, and very uncomfortable. In this sense, ride comfort becomes an indicator of the track's health which other parameters are not able to recognize at the level or irregularities observed on the track[8].

4.6 Maintenance activities to reestablish the track geometry

The track geometry is an important factor for the stability of the track and the vehicle response. Overtime the ballast shifts and settles, causing the track to lose its original geometrical configuration (e.g. cross section). To reestablish the track geometry, prevent its rapid deterioration, and adverse vehicle responses, specific maintenance activities can be performed. Activities to reestablish the geometry of the track include tamping, stabilization, stone blow (used in Great Britain), cleaning (to increase friction between rocks – increasing loading capacity of the ballast), as well as drainage improvements to prevent water accumulation (wet spots, track softening) (Tzanakakis 2013; Wenty 2007; Lichtberger 2007a).

Tamping is performed to adjust the vertical profile and lateral geometry of rails (Li et al. 2016). It restores the track position by providing equal supporting piles of compacted ballasts under the sleeper which should display a homogenous vertical stiffness and load distribution (Lichtberger 2005a). An improvement of standard deviations of a track's parameter due to tamping can be seen in Figure 48.

To implement tamping, a machine lifts the track to a desired level, creating a void between the bottom of the sleepers and the top of the ballast (in place). The machine inserts tines into the ballast on either side of the sleeper. The tines are squeezed to move the ballast into the void (Li et al. 2016). To perform an optimum tamping, and avoid damage to the ballast during the tamping operation, there are typical working parameters that need to be met (Lichtberger 2005a). These parameters are namely, the frequency should be 35 Hz, the tine amplitude 1 – 5 mm, the pressure 115 – 125 bar, the squeezing time 0.2 – 0.8 s and the tamping depth 15 – 20 mm, which is the free space between the bottom of the sleeper and the upper edge of the tamping tine plate.

---

[8] Higher irregularities, such as the synthetic irregularities produced in this study, might present a different situation and the other parameters, namely safety or pressure on the subgrade due to high wheel forces, would need to be evaluated accordingly as performed in section 5.3.

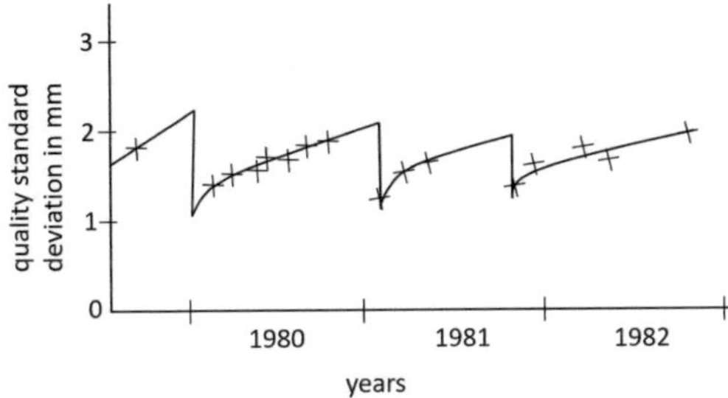

Figure 48: **Improvement of standard deviations of a track's parameter due to tamping per (Lichtberger 2005a)**

In addition to reestablishing the geometrical configuration of the track, tamping has an equalizing effect and establishes a favorable rate of deterioration (Lichtberger 2005a).

Even though tamping generates an excellent track geometry (Wenty 2007), it reduces the ballast unit weight (e.g. through abrasion) and bulk density for the depth of insertion (Li et al. 2016). The reduction of bulk density has been reported to correspond to a decrease in track lateral resistance of up to 40-60%, which might lead to rail buckling (Wenty 2007; Lichtberger 2007b). Track stability is regained either through a mechanized stabilization process, equating to 100,000 tons of traffic (Lichtberger 2007b), or through a process called initial settlement of the track occurring after about 0.5- 2 million gross tons (MGT) of passing trains (Lichtberger 2005b).

The track mechanized stabilization plays an important role in the reestablishment of a track geometry that provides stability and slower rates of deterioration. Mechanical stabilization is introduced through machines which grab the rails with roller clamps while applying a horizontal oscillation under a steady vertical load (Lichtberger 2007b). The horizontal and vertical motions have shown to provide a more controlled track consolidation and even settlement than the passing of trains (Lichtberger 2007b). The mechanized stabilization process produces a reduction in the loss of geometrical quality during the initial MGT (Lichtberger 2001; Lichtberger 2005b). An additional benefit of the mechanized process is that no speed restrictions are imposed on the track during the settlement period, which reduces operational hindrances and hence costs (Lichtberger 2005b).

The first phase of stabilization of a track under traffic is characterized by an exponential and rapid settlement (Lichtberger 2005a) which as mentioned before is uneven due to the stochastic behavior of the forces produced by the vehicles (Lichtberger 2007b). To compensate for this rapid settlement, a tamping method called design over-lift is used which provides a smooth profile which remains smooth in the subsequent settlement which takes place at a slower rate (Li et al. 2016). In general, over-lift seeks to overcome the so-called "track memory" phenomenon, which is the quick return of the track to almost the same defective position in which it was before tamping (Li et al. 2016). Over-lifting performance can be further improved with mechanized stabilization (Lichtberger 2005a).

However, once the ballast stones wear out due to splitting and abrasion, and the ballast bed becomes dirty or fouled, the tamping and stabilizing actions are no longer sufficient to generate a good track geometry and their positive effect on the geometry and deterioration rate is not long lasting (Lichtberger 2007b; Schilling 2005; Lichtberger 2007b). Furthermore, fouled ballast leads to an accelerated destruction of the ballast and increases the deterioration rate of the track (Lichtberger 2001). Additionally, fouled ballast causes a loss of carrying and load distribution capacity allowing higher pressures under the sleepers and subgrade (Lichtberger 2007b) which compromises the stability of the track. As explained in (Schilling 2005), to restore the ballast bed to its original condition, the ballast needs to be cleaned or completely replaced. Water logging or stagnation goes hand to hand with fouling of the ballast bed. It is essential for a healthy ballast to promote good drainage (Powrie and Louis 2016). Cleaning of the ballast, across the whole cross section of the track, including the shoulder of the track section, enables good drainage (Lichtberger 2007b), which prevents the development of track fouling.

4.7 Service life of track components and time of track intervention

The track is a complex system whose interaction needs to be analyzed in an integral manner. This is no exception to the maintenance phase of the track's life cycle where different components present different life cycles. In (Lichtberger 2007b), typical values of life cycles for track components and maintenance activities were presented for a typical intensively operated main line. In general, tamping (stabilization) should be performed every 40 to 70 million tons or every 4 – 5 years while ballast cleaning every 150 – 300 million tons or 12 -15 years, which should be undertaken only in combination of other major work, for example sleeper renewal. The replacement of ballast should be every 200 – 500 million tons or 20 – 30 years and the rail renewal is suggested every 300 – 1000 million tons or 10 – 15 years. In LRT systems the service life of concrete sleepers has been reported to be 50 years under certain design practices and high-quality wooden sleepers display a service life of 20 – 30 years.

The challenge lays on performing the right combination of actions to maximize the service life of the entire track for the required track quality. Research have shown that combination of actions to establish the condition of the components, have beneficial or detrimental effects on the track's performance and service life. For example, (Lichtberger 2005a) shows that a combination of tamping and rail grinding, immediately after, controls the increase of dynamic stresses that normally appear after tamping and that result in a higher deterioration of the track geometry. The latter was also pointed out in (Popovic et al. 2017) which shows a synchronization of maintenance activities seeking to coordinate the replacement of the track components known as the principle of integrated maintenance as seen in Figure 49. The approach consists of the coordination of cyclic grinding, tamping cycle, renewal of rails/sleepers and ballast replacement/cleaning.

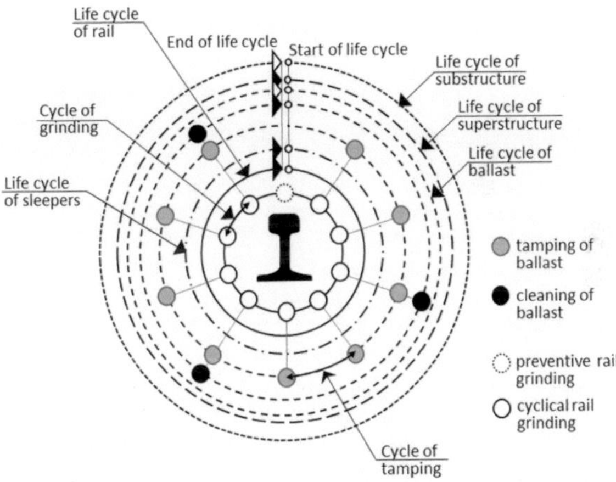

Figure 49:　　　**Principles of an integrated track maintenance per (Popovic et al. 2017)**

5 Determination of an Intervention Limit

The importance of the determination of limits for maintenance and renewal in light rail systems has been established throughout this document. In this chapter, limits are established using all aspects and processes previously laid out (e.g. data treatment) as well as through MBS simulations using SIMPACK©. Simulations are performed using a vehicle model representing the vehicle operated on the infrastructure in Stuttgart, Germany (i.e. SSB DT8). The model used was provided by the Institute of Machine Components (Ger. acronym IMA) of the University of Stuttgart to conduct track/vehicle collaboration studies.

5.1 Multibody simulation

MBS is a numerical method to simulate the mechanical behavior of multiple bodies (mechanical components) interconnected by massless kinematic joints and constraints, both of which control or restrict the relative motion of the multibody components (Dassault Systems Simulia Corp. 2017; Flores 2015). From the relative motion of the bodies, force elements generate applied forces and torques, typically represented by springs, dampers, and actuators combined, for example, into primary and secondary suspensions (Schupp et al. 2004). In contrast, joints give rise to constraint forces by restricting the relative motion of the bodies (Schupp et al. 2004). MSB is used to study the kinematics and dynamics characteristics of systems. The target is to represent reality in order to obtain information regarding the mechanical behavior of the vehicle without complex measurement equipment or expensive test runs. MBS can be used for the evaluation of comfort, safety and performance of a vehicle.

To build a model that represents an abstraction of reality, it is necessary to define the minimum individual bodies needed and the type and number of the connection elements (Holzscheiter 2017) that reflect their proper contribution to the system (Garg and Dukkipati 1984). Bodies might be flexible or rigid; containing mass and inertial properties (Rill and Schaeffer 2010). Rigid bodies are those which cannot deform but only rotate and translate, while flexible bodies are elastic structures with internal dynamics which are able to deform (Flores 2015). Bodies in railway applications are significantly stiff (Garg and Dukkipati 1984) and can be considered to be perfectly rigid so their flexibility can be disregarded (Flores 2015). This assumption simplifies the process of modeling of multibody systems, e.g. flexible bodies require significant longer calculation times (Rill and Schaeffer 2010).

Bodies (n number of bodies) are located in space and oriented in relation to one another and to a common reference frame (i.e. inertial system) (Flores 2015). Connections are achieved through global and local coordinate systems (Dassault Systems Simulia Corp. 2017) which at the same time join force elements to bodies. Global coordinates are represented by three orthogonal axes, $x_0 y_0 z_0$, that are rigidly connected to a point called origin (Flores 2015). Local coordinates, $x_n y_n z_n$, define local properties of points that belong to a body and are generally at the bodies' center of mass (COM). Local coordinate systems translate and rotate with the body motion, consequently, their position and rotation vary with time (Flores 2015). Figure 50 shows two rigid bodies whose positions and orientation are described through global and local coordinate systems.

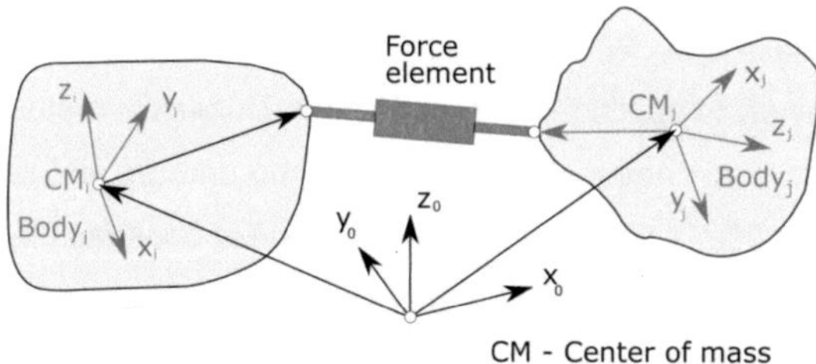

Figure 50:   **Global and local coordinate systems showing the relationship of two rigid bodies to one another and to a global reference adapted from (Rill and Schaeffer 2010)**

Models have a minimum degrees of freedom (DOF) describing the system. This is represented by state variables defining the relative motion of the bodies. Joints, flexible bodies and dynamic states provide one DOF while each constraint state removes one DOF by constraining movements in a particular direction (Dassault Systems Simulia Corp. 2017; Flores 2015).

The behavior of an MBS is described by the equations of motion formulated through coordinates, generally categorized as dependent or independent (Flores 2015). Independent coordinates are free to vary arbitrarily, while dependent coordinates are required to satisfy the equations of constraints. The process of determining the equations of motion is the modelling part; following two general approaches: point to point coordinates (see Figure 50) and body coordinates. The first one assigns coordinates to joints and constraints to bodies while the second does the opposite (Flores 2015). According to (Flores 2015), body coordinates are more convenient for the analysis of multibody systems and their formulation is a systematic approach to generate the equations of motion based on the Newton-Euler equations.

An important aspect in the modelling of an MBS is the wheel-rail normal contact. This interaction is often approximated as a Hertzian spring (Bezin et al. 2009) describing the contact patch (rolling contact) formed by the slight deformation of the wheel and the rail resembling an ellipse of minute surface area (Hertzian Theory). According to contact mechanics theory (i.e. Kalker's theory), the elliptical curve of the contact patch is characterized by two areas, namely the stick and slip areas

As explained in (Rill and Schaeffer 2010), the wheel-rail contact patch and pressure distribution are calculated using the Hertzian theory. The pressure distribution in the contact patch results in a normal contact force exerted by the vehicle which in conjunction with the coefficient of friction results on tangential forces F_t1 (shear traction) and F_t2 as well as the moment force M_b (Rill and Schaeffer 2010) as seen in Figure 51.

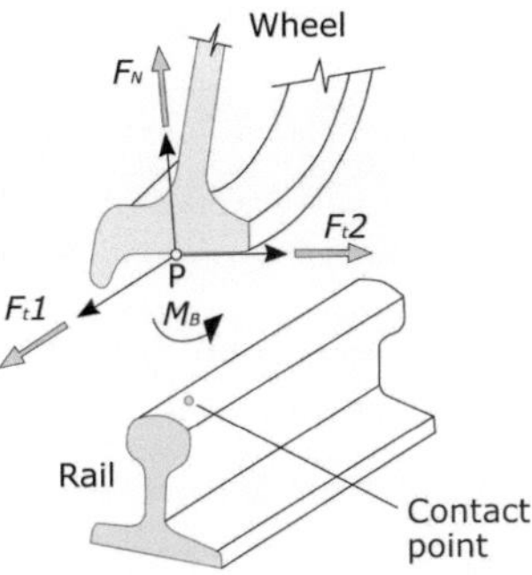

Figure 51: **Tangential forces and moment developed at the contact patch adapted from (Rill and Schaeffer 2010)**

To determine the tangential rail-to-wheel contact forces, the MBS software SIMPACK© uses the FASTSIM method ("A Fast Algorithm for the Simplified Non-Linear Theory of Contact") (KALKER 1982; Dassault Systems Simulia Corp. 2017). The method consists of the discretization of the contact patch into slices parallel to the x-axis (KALKER 1982). The tangential forces are determined by a simplified numerical integration of the tangential stresses in the slice elements (Dassault Systems Simulia Corp. 2017;

KALKER 1982). As mentioned in (Bezin et al. 2009) the discretization is performed in order to account for the geometrical non-linearities within the contact patch.

In SIMAPCK© the track is represented by a movable track with a movable force element that models the track's elasticity with a linear stiffness and damping (Dassault Systems Simulia Corp. 2017) on a spring and damper model. This type of model reduces the number of degrees of freedom of the system and decreases simulation time (Bezin et al. 2009). The ballasted track is modelled using three masses: the left and right rails, and the sleeper–ballast combined mass (Bezin et al. 2009; Dassault Systems Simulia Corp. 2017).

The track modulus in the simulation software is represented by force elements containing stiffness and damping values. Per section 4.1, the track is considered to have a good structural quality whose stiffness and damping values, in the vertical direction, are calculated following approaches in (Göbel and Lieberenz 2004; Vogel et al. 2015) (see Appendix VI: Stiffness and damping calculations) and presented in Table 21.

Track Modulus [MN/m^3]	Track Structural Quality	Damping [kNs/m]	Track Stiffness [kN/m]
50	Poor	277	250
100	Good	327	480
150	Very Good	375	940

Table 21: **Track modulus, damping and stiffness defining the track structural quality of the track (own work)**

5.1.1 Light Rail Vehicle (LRV) model

This research uses the MBS model of an LRV from the Stuttgart Tramway Association (Stuttgarter Straßenbahnen AG, SSB) named DT8.10 (Figure 52).

Figure 52: **DT8.10 LRT vehicle used to model MBS (photo: author)**

The DT8.10 model has been used in several collaborative studies between the Institute of Railway and Transportation Engineering (Ger. acronym IEV) and the Institute of

Machine Components (IMA) of the University of Stuttgart (Skorsetz et al. September 23 -27; Strobel et al.) and studies conducted by IMA (Holzscheiter 2017). The vehicle consists of two car-bodies (twin unit) coupled together permanently. The vehicle has eight powered wheelsets in four bogies as seen in Figure 53.

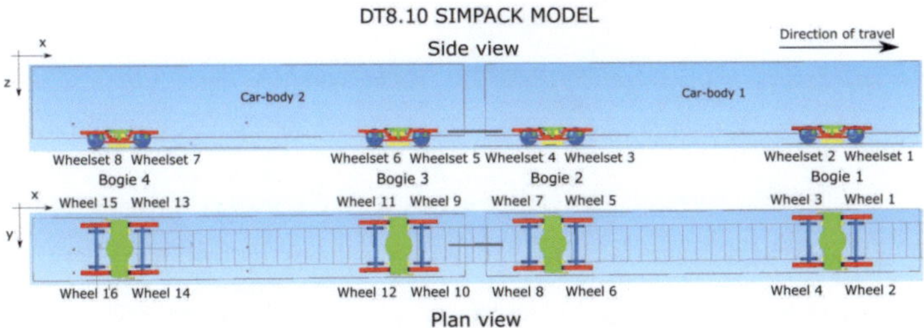

Figure 53:   **DT8.10 SIMPACK© model (own work based on screenshot of MBS model)**

Car-bodies pivot on a bearing seating on the bolster linked to the bogie frame via a secondary suspension. A bolster-stop controls lateral movements of the car body. The secondary suspension consists on air springs in parallel to hydraulic (oil based) dampers. The system also includes a primary suspension which consists of compression springs acting in parallel to hydraulic (oil based) dampers (Holzscheiter 2017). Wheelsets, bogie frames, car-bodies and other components are represented as rigid bodies, connected by spring-damper force elements, joints or constraints. The model consists of 43 rigid bodies connected by 162 springs and dampers and forms a "swinging-like" system of 120 DOF. Figure 54 depicts a schematic view of the model.

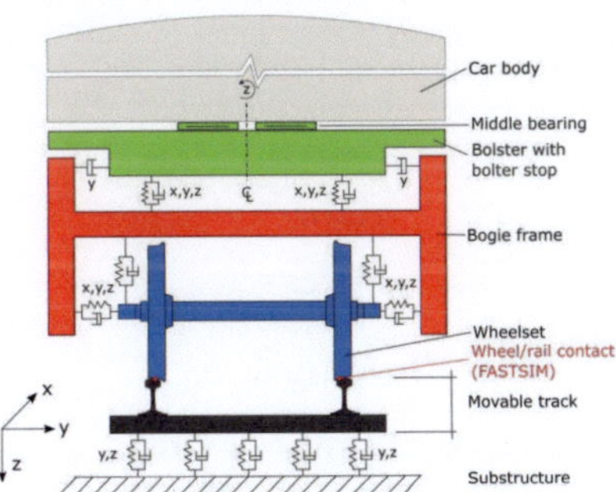

Figure 54:   **Schematic configuration of simulation model adapted from (Strobel et al.)**

As seen above, the vehicle's suspension is modelled using two spring levels, namely, a primary suspension between wheelsets and bogie frames and a secondary suspension between bogie frames and car-bodies as seen in Figure 55.

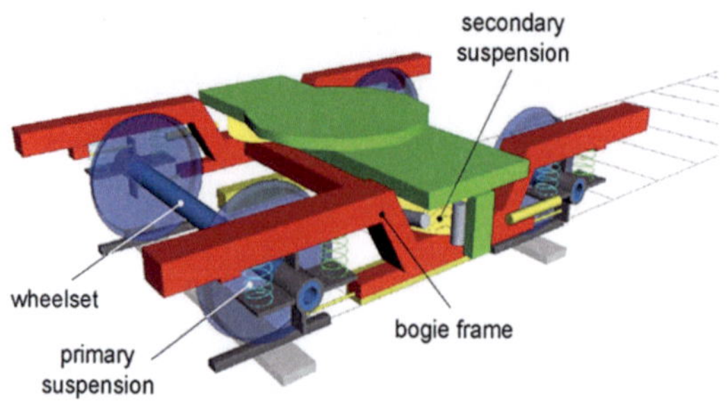

Figure 55:   **SIMPACK© model of bogie and suspensions per (Skorsetz et al. September 23 -27)**

To obtain a model representing reality, all important data and properties of the original LRV were provided by SSB: mass properties, vehicle dimensions, stiffness of springs, damping rates (see Appendix VII: Vehicle Data) and specific SSB wheel profile. Thus, it is possible to build a model displaying similar vibration performance as the original vehicle (Skorsetz et al. September 23 -27). For example, acceleration measurements were taken in the cabin of a real vehicle running at varying speeds (e.g. average speed 44.5 km/h in the run presented below) and then compared to accelerations obtained through simulation at 50 km/h for day 1756 which corresponds to a date close to the days the accelerations were taken (see Figure 56).

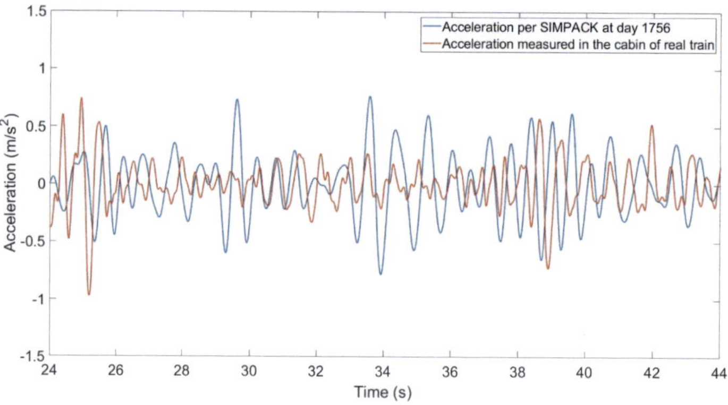

Figure 56:   **Comparison of MBS model and real train accelerations (own work)**

As it can be seen in Figure 56, the accelerations from simulations are higher in some locations than the accelerations measured in the vehicle. However, both are still within a reasonable range from one another (see Appendix IX: Real vs Simulated Vertical

Accelerations for more comparison with real measurements). The slight difference observed might be due to several reasons: a) the varying and lower average speed of the vehicle for which the acceleration is being measured, b) the vehicle where measurements were taken might not have exactly the same suspension conditions as the vehicle modeled and c) the accelerometer used is not completely appropriate for the measurements[9] or even d) changes in track stiffness, although (Frohling et al. 1996) mentions that through the analysis of vehicle responses it was found that differences in the RMS values of a vertical body acceleration due to changes in track stiffness was small compared to dynamic loading.

Regarding the track, it was modeled as a straight segment using an 49E1 (S49) rail profile on a standard gauge (i.e.1435 mm). Track irregularities deriving from measured data are added to the track via track excitations as explained in section 5.1.2.

To measure train accelerations, six sensors were placed in the passenger cabin at representative positions according to standard EN 12299 (The British Standards Institution 2009) and two more in the axle box of wheels 15 and 16 (see Appendix VIII: Location of Sensors). The sensors implemented measure accelerations in x-,y- and z-direction during simulation. The data generated in the cabin are used in post-processing to assess the comfort level following the method presented in section 2.4.5. The data measured at the axle box are filtered appropriately based on provisions presented in (The British Standards Institution 2016) and provide the possibility to obtain track irregularities in the vertical direction through double integration as shown in section 5.2.

As seen in Figure 57, an interesting aspect of measured accelerations in the passenger's cabin is that the signals for the six sensors appear close to each other.

---

[9] Data gathered with a mobile phone with an accelerometer and gyro system. The application used is called spark, which measures accelerations at a rate of 200 Hz (frequency also used in the MBS program). The data measured was treated with a band pass filter with cutoff frequencies between 0.4 – 3 Hz resulting from an FFT analysis in MATLAB that reveals the frequencies of interest.

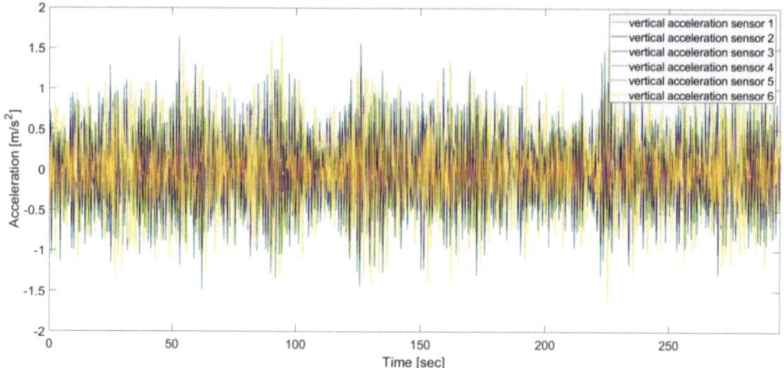

Figure 57:   **Accelerations measured by six different sensors spread in passenger cabin according to (The British Standards Institution 2009) (own work)**

However, when the RMS of the vertical acceleration (see Figure 58) and comfort level for accelerations in all directions (see Figure 59) are calculated, greater and more significant differences between signals can be observed. As seen in Figure 58 and Figure 59, sensors 1 and 6 display the higher values. This was observed in all runs / simulations performed. This observation is important to consider because for this study the worst vehicle reactions (higher accelerations) are the most important in the determination of the limits of intervention (i.e. an average value of sensor 1 and 6).

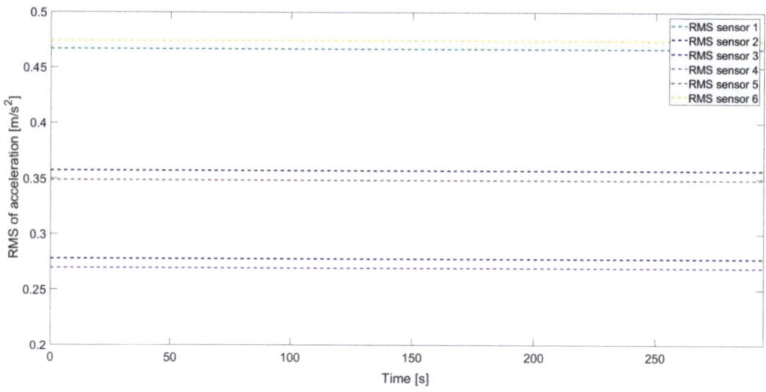

Figure 58:   **RMS of signals measured by six different sensors within the passenger cabin per (The British Standards Institution 2009) (own work)**

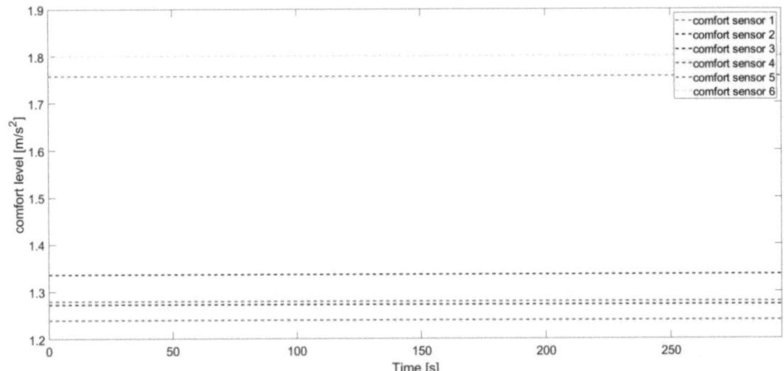

Figure 59: **Comfort level of signals measured by six different sensors within the passenger cabin calculated per (The British Standards Institution 2009) (own work)**

5.1.2 Excitations in MBS model

In SIMPACK© excitations (i.e. track irregularities) describe superimposed irregularities to an ideal track layout (Dassault Systems Simulia Corp. 2017). Excitations are divided into track related and rail related. Track related refer to the average of both rails in the vertical, lateral directions in addition to the roll (cross level) and gauge. Rail related excitations refer to irregularities in the lateral, vertical and cross level directions for each rail separately. The data provided by SSB includes rail and track related irregularities. Since vertical irregularities were only provided for track related excitations, the track related option in the MBS model was chosen. Horizontal left and right rail irregularities were averaged according to the formulas presented in section 2.3.1 to convert them to track excitations.

Three basic methods are used in SIMPACK© to create or assign excitation to a track, namely harmonic functions, input functions (e.g. measured track irregularities) and pseudo-stochastically from PSD standards. The PSD method assigns randomly generated phases to take care of the lack of phase information of PSD signals.

This research evaluates vehicle running characteristics and responses related to ride quality. As mentioned in section 2.4.5, one aspect to be assessed is comfort which according to (The British Standards Institution 2009) requires track data equivalent to a five-minute ride for the speeds investigated (i.e. between 10 – 80 km/h). However, data corresponding to five-minute rides at the evaluated speeds are in general not available for LRT systems. This is due to the distance between stops (e.g. 500 m) which is much shorter than the distance required for simulation (Strobel et al.) (i.e.

continuous data should be between 830 and 6,600 m for the speeds simulated). Therefore, real measured data cannot be used directly in simulations and needs to be processed for its use in the software.

To model longer track segments displaying similar track geometry properties as the real data, a PSD function is used which maintains the statistical and frequency characteristics of real irregularities and can be in turn converted back to spatial domain signals. The transformation from frequency to spatial domain is performed through an inverse transformation operation (i.e. similar to an inverse Fourier transformation). The obtained spatial signals can then be applied to the length of track needed for the five-minute vehicle run required by the comfort evaluation. To do this the PSD data of measured data (see section 3.3.2) is imported into SIMPACK©. Figure 60 describes the process to generate excitations in SIMPACK©.

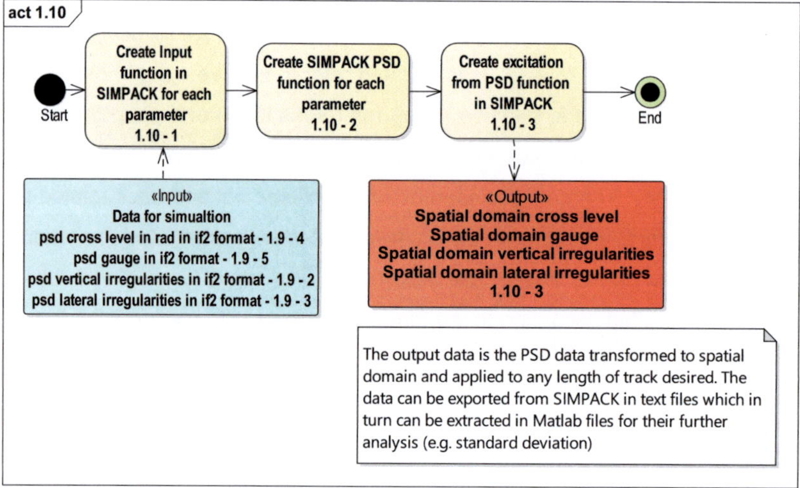

Figure 60: Process for excitation creation in SIMPACK© based on PSD functions of measured track irregularities (own work)

Figure 61 shows the comparison of a signal measured on the track against a synthetic signal created in SIMPACK© (labeled in Figure 60 as 1.10 -2).

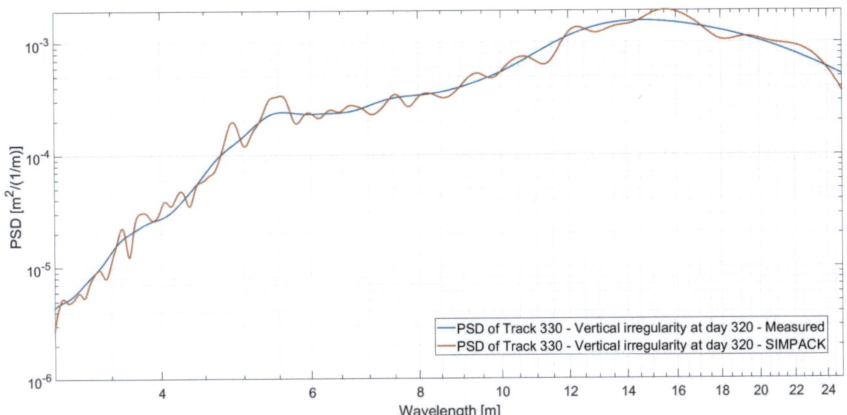

Figure 61: Comparison of PSD of signal day 320 as measured and as obtained in SIMPACK© (own work)

In consequence, as it can be seen in Figure 62, the spatial signal generated in SIM-PACK© (labeled as 1.10 – 3 in Figure 60) displays similar spatial characteristics as the measured signal.

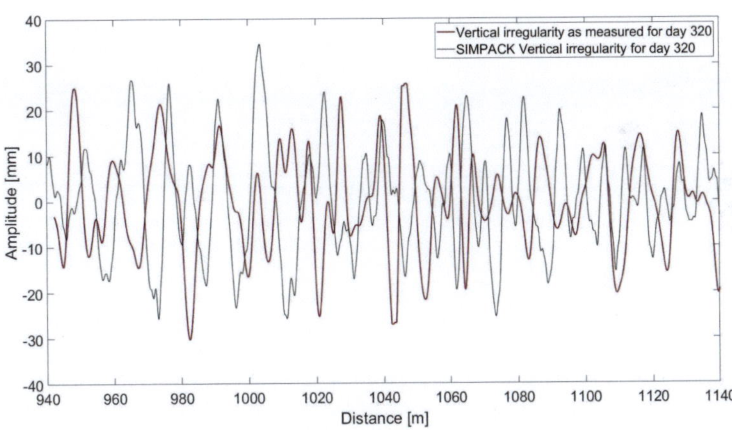

Figure 62: Comparison of signals generated in SIMPACK© through PSD functions of measured signals vs the signal measured in spatial domain (own work)

5.2 Simulation – vehicle responses

Simulations were performed for speeds between 10 – 80 km/h to determine the vehicle riding characteristics, namely vertical load imposed on the track (Q), derailment coefficient (Y/Q) and comfort level. The value of Y/Q obtained is pretreated and evaluated against the limit values as established in Table 4 section 2.4.4. The vertical load is pretreated according to section 2.4.4, then the stress on the subgrade σ_z is calculated (see Appendix X: Vertical Stress Calculation) and compared to the allowable stress σ_{allow} at depths of 50 and 60 cm (representative depths in ballasted track construction). The comfort level is calculated from accelerations measured in the cabin and compared to the limits established in (The British Standards Institution 2009). If the limit of vertical load, converted to a maximum stress at the subgrade, is reached, it represents a limit to the deterioration of the track caused by the vehicle. If the derailment coefficient is reached, it represents a safety limit. As mentioned before, the comfort index is taken as an indicator of track deterioration and degradation process. The limit values used for evaluation are shown in Table 22.

	Limit value for comfortable level [-]	Limit value for half distance to uncomfortable level [-]	Limit value for uncomfortable level [-]	Limit value for very uncomfortable level [-]
Mean Comfort Index N_{MV}	2.5[10]	3.0[11]	3.5[11]	4.5[11]
	Limit value for safety [-]			
Derailment coefficient Y/Q	0.8 (1.2[12])			
	Allowable pressure [N/cm^2]			
Stress σ_z at 50 cm	2.46[13]			
Stress σ_z at 60 cm				

Table 22: **Limits for mean comfort, derrailment coefficient and allowable subgrade stress (own work)**

However, to establish limits of intervention, the parameters need to be compared simultaneously to one another to determine which one represents the better measure of track deterioration and degradation (i.e. the one reaches its limit before the others) for the worst quality recorded (e.g. for track 330 at day 320). A comparison at different speeds is done for this purpose. The evaluation of the forces on the track and derailment coefficient is obtained for every wheel (16 wheels in this case study) of the vehicle to establish the highest possible value (see Table 23).

As seen in Table 23, the vertical load (converted to a stress at the subgrade) and the derailment coefficient do not reach or get close to their respective limits (red values)

---

[10] Values based on table 8 of standard 12299 [The British Standards Institution (2009)].

[11] The limit for half way to uncomfortable was taken as a warning limit at this stage of the research. The value was obtained from the value between the comfortable level and uncomfortable level.

[12] Absolute maximum allowed per standard 14363:2006 table 4 comment a [The British Standards Institution (2016)]. In this study, upon reaching this value, the track is suggested to be closed for operation.

[13] Value calculated as shown in Appendix X: Vertical Stress Calculation.

even at the highest speed (i.e. 80 km/h). Opposite to that, the comfort index is close to the uncomfortable level at 70 km/h (value in yellow) and by 80 km/h is well over the uncomfortable limit (red value). Up to this point (worst track geometry observed), the mean comfort index provides a better clue of the deterioration of the track in consideration to the LRV responses.

Speed [km/h]	Comfort Index [-]	Comfort index limits	Y/Q [-]	Y/Q limit [-]	Q [kN]	Q_{max} [kN]	σ_z [N/cm^2] 50 [cm]	60 [cm]	σ_{allow} [N/cm^2]
10	0.09		0.1370		41.74	52.07	1.11	0.85	
20	0.38	2.5	0.1670		42.30	52.77	1.12	0.86	
30	0.80	3.0	0.1535	0.8	42.52	53.04	1.13	0.87	
40	1.23		0.1503		43.10	53.77	1.15	0.88	2.46
50	1.78	3.5	0.1580	1.2	44.90	56.01	1.20	0.92	
60	2.45		0.1633		46.70	58.60	1.26	0.96	
70	3.38	4.5[14]	0.1903		49.43	61.98	1.33	1.01	
80	4.00		0.2338		52.13	65.71	1.41	1.08	

Table 23:   **Comparison to establish most suitable parameter to determine limits of intervention (own work)**

Since comfort demonstrates to be a good measure of deterioration, simulations were performed for all inspection dates available to establish the development of the comfort deterioration (see Figure 63).

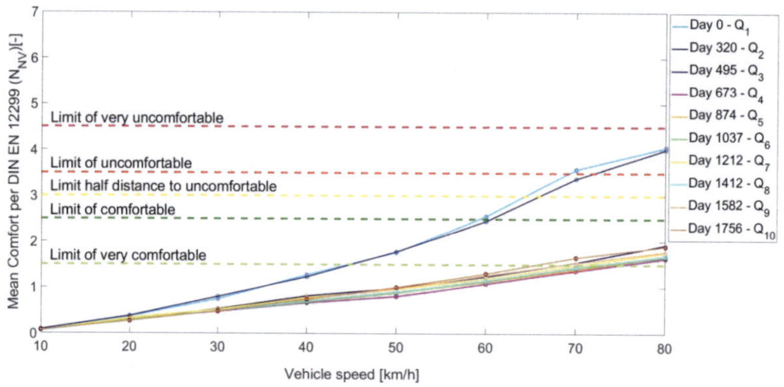

Figure 63:   **Mean comfort index for all days plus day 320 with increased magnitudes (own work)**

---

[14] Limit of very uncomfortable per BS EN 12299 [The British Standards Institution (2009)]

As it can be seen, it is evident that days 0 and 320 show a higher level of deterioration. In fact, as mentioned above, at speed 70 km/h and 80 km/h the track shows a level of comfort requiring attention. Notwithstanding, track 330 is not driven at such higher speeds. In fact, the track is mostly driven closer to 50 km/h. At this speed the vehicle still experiences a comfortable ride and is well under the maximum value of the derailment coefficient. Furthermore, even when at 60 km/h the comfort was still acceptable, the track was renewed. In this sense, if renewal actions were purely related to vehicle responses[15] (and track segmentation done in speed categories), the renewal would seem to have been done rather early.

A closer look at the development of the comfort index (Figure 64) shows that day 673 (fuchsia line) displays the best quality after the renewal of the track. From this point on, the track starts its deterioration process. Likewise, it can also be noticed that the more substantial difference in comfort index occurs between each inspection date above 40 km/h. This behavior agrees with Figure 7, which shows that signals with a wavelength content between 10 and 25 meters exert a bigger influence on the vehicle responses at speeds higher than 40 km/h.

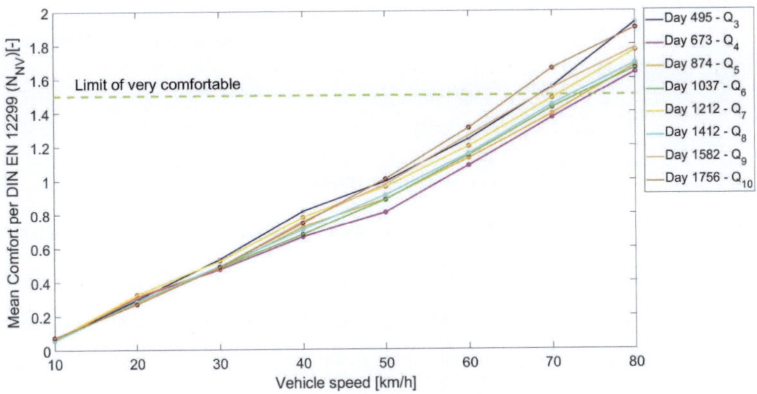

Figure 64:   **Mean comfort index development for days after renewal (own work)**

Furthermore, the deterioration of the comfort index agrees with the changes in standard deviation observed for vertical and horizontal irregularities from day 673 on (see

---

[15] Renewal could be actually be performed due to track component aging, e.g. end of service life of sleepers.

Figure 65). Similarly, it agrees with the track characterization studies performed using spatial – frequency domain and fractal methods as shown in (Álvarez 2018).

Also, it is important to point out, that the alert limit prescribed in standard EN13848-5 is surpassed by vertical irregularities for all inspection days, while the lateral irregularities never exceed their limit; even before the renewal of the track. This suggests that vertical irregularities deteriorate more than lateral irregularities (which agrees with (Álvarez 2018)). It also reinforces the notion that conventional railway standards, at least for the vertical irregularities, might be too strict for this system. Moreover, a higher level of irregularities in the vertical direction pose a higher influence on the geometrical quality, and therefore, on the responses of the vehicle.

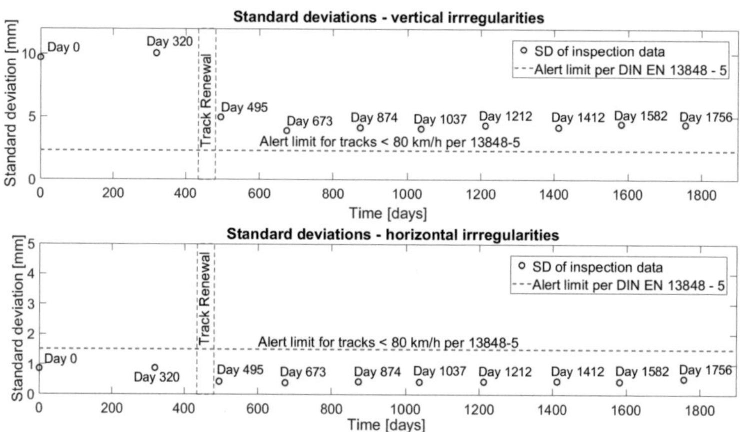

Figure 65: SD of vertical (top) and horizontal (bottom) irregularities for track 330 (own work)

Vehicle responses produced on the axle box of the vehicle provide an additional way to verify whether the magnitude of the accelerations produced through simulations are sensible. This is done through double integration of the axle box accelerations in the $\ddot{y}_{axle}$ and $\ddot{z}_{axle}$ directions (Table 24). The latter should correspond to the vertical irregularities of the track as shown in Figure 66.

$\ddot{y}_{axle}$	Lateral axle box acceleration
$\ddot{z}_{axle}$	Vertical axle box acceleration

Table 24: Axle -box accelerations to obtain track irregularties from double integration

To perform a double integration analysis, vertical accelerations in the axle box are treated before evaluation per the recommendations made in Table 5 in standard EN

14363: 2016 (The British Standards Institution 2016). The assessment parameters for riding characteristics are shown in the Table 25 along with their respective evaluation filters and percentiles.

Assessment parameter	Symbol	Unit	Evaluation filter	Percentile
Acceleration in vehicle body	$\ddot{y}^*_{q,max}$ $\ddot{z}^*_{q,max}$	m/s^2	Band pass filter 0.4 to 10 Hz	h1 = 0.15% h2 = 99,85 %

Table 25:   **Assesment parameters for riding characteristics per (The British Standards Institution 2016)**

However, according to (Andrea Haigermoser 2013) acceleration signals are filtered with a high pass filter with a cutoff frequency of 0.4 Hz. Nonetheless, the filter cut-off frequency can be established by performing a Fast Fourier Transform (FFT) analysis on the acceleration data. This was the method used in this study. Figure 66 shows the double integration of the vertical axle box accelerations obtained in simulation compared to the measured and treated vertical irregularities and the vertical irregularities used in simulation for day 320. As it can be seen, the double integrated simulated accelerations produced a similar signal as both, the simulated and measured vertical irregularities. This shows that simulations produced coherent results in relation to the measured data provided.

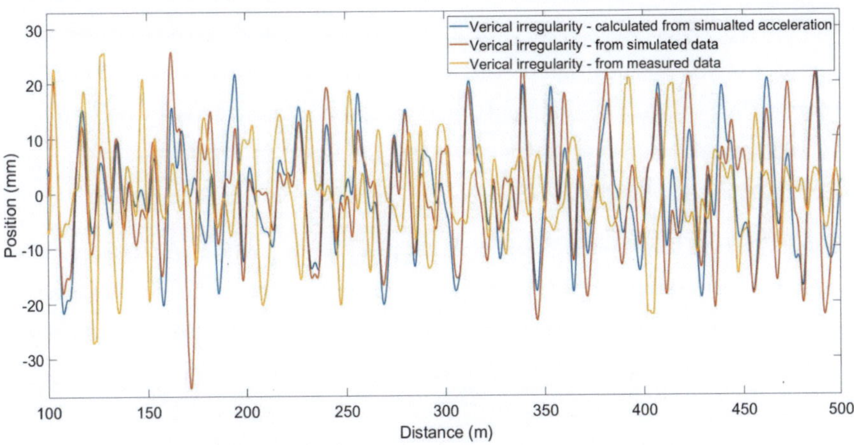

Figure 66:   **Comparison of vertical irregularities produced through double integration versus measured (provided) and simulated vertical irregularities (own work)**

5.3 Limit definition at 50 km/h

Since simulations in section 5.2 already represent the worst geometrical quality observed on track 330 and no adverse vehicle reactions were observed at 50 km/h, it was necessary to generate synthetic signals that contained higher track geometrical irregularities. However, it was also necessary to keep the characteristics of track irregularities observed in the infrastructure under study. The synthetic signals with higher irregularities and geometrical characteristics were achieved in the frequency domain by increasing the magnitude of the PSD of all parameters for the worst TGQ observed for track 400 (see section 5.2 and Appendix XI: Track Parameter Magnitude Increase). It was expected that at 50 km/h[16] the synthetic irregularities would cause the vehicle to either reach the limits of comfort, subgrade vertical stress, or derailment coefficients (i.e. 0.8 and 1.2).

First, the synthetic signals were assessed through the effects of each parameter in terms of the comfort level, which had been determined to be the most sensible indicator of track deterioration. In this sense, a study found that vertical and lateral irregularities have the largest effect on comfort level (Araji 2018). Furthermore, (Araji 2018) suggested a process to find the appropriate factor increase. The process was iterative, assuming that the worst day on track 400 corresponded to a factor 1 from which a gradual factor increase was performed (see Appendix XI: Track Parameter Magnitude Increase). The result shows that it is acceptable to increase all parameters by the same factor at the time; increased until, through simulations, the limits of comfort are reached. The factor increase study agrees with conclusions of track characterization studies performed by (Álvarez 2018) through fractal and spatial and frequency domain methods, which determined that vertical irregularity deterioration dominates in the infrastructure analyzed (Álvarez 2018).

Figure 67 shows the effect of the higher irregularities on comfort. As it can be seen, increasing the magnitude of irregularities by a factor of 2, the limit of comfort is almost reached. Likewise, limits of uncomfortable and very uncomfortable were reached or almost reached at a factor of 5 and 9 respectively.

---

[16] Representative moving speed observed through test runs on track 400.

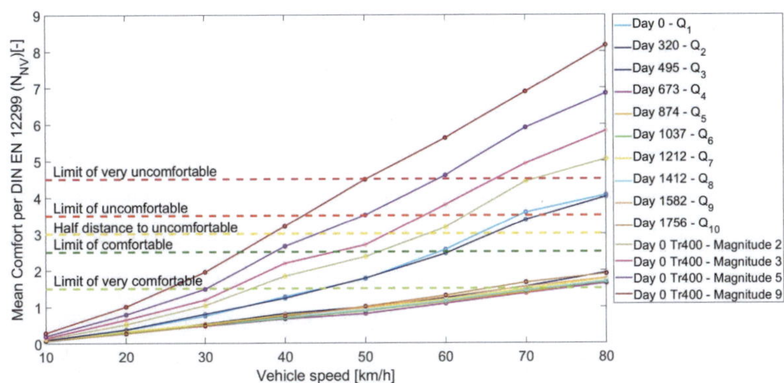

Figure 67: **Comfort index for synthetic signals displaying higher track irregularities (own work)**

However, increasing irregularities beyond the values presented in Table 23 means that safety and track loading/allowable pressure at subgrade might be compromised. Table 26 shows results of 10 simulations per increased factor at 50 km/h.

In Table 26 it can be seen that not just the comfort level has been surpassed in all instances, reaching even almost a very uncomfortable limit by factor 9, but also at this point, the safety level is compromised.

Track / Day	Factor	Comfort Index [-]	Comfort index limits	Y/Q [-]	Y/Q limit [-]	Q [kN]	Q_{max} [kN]	σ_z [N/cm²] 50 [cm]	σ_z [N/cm²] 60 [cm]	σ_{allow} [N/cm²]
400/0	2	2.31		0.27		46.72	58.28	1.25	0.95	
400/0	3	2.75	2.5 3.0 3.5	0.38	0.8 1.2	48.39	60.37	1.29	0.99	2.46
400/0	5	3.47	4.5[17]	0.55		51.07	63.72	1.36	1.04	
400/0	9	4.42		1.49		53.65	66.63	1.42	1.09	

Table 26: **Factor increase to determine limit values (own work)**

---

[17] Limit of very uncomfortable per BS EN 12299 [The British Standards Institution (2009)].

The first factor-increase approach determined rough magnitudes in consideration of vehicle responses; however, "fine-tuning" to obtain the exact values required. This was achieved by refining the values of the increasing factors and performing ten simulations per factor increased (i.e. 280 simulations for an average of 1hr 30 min each on a 2.10GHz processor with 32 GB of RAM). Table 45 in Appendix XII: Magnitude Increase "Fine-Tuning" shows the factors simulated to achieve values closer to the expected limits. To visualize the limit values in regard to vehicle responses, the values where normalized to a maximum value (see Figure 68). The maximum value was not taken to be the worst possible value allowed. For example, regarding comfort, the worst value was not represented by very uncomfortable but by uncomfortable. This is because it cannot be expected that an infrastructure is allowed to reach such a state. Likewise, from simulations it was observed that the derailment coefficient exceeded the maximum value of 0.8 established in standard 14363, which according to the same standard might be acceptable for an absolute maximum value of 1.2. However, it would not be acceptable to allow an infrastructure to achieve that level of deterioration. Hence, the normalized values were allowed to be greater than 1 namely 1.3 for the comfort level (i.e. comfort level 4.5 for a very uncomfortable ride) and 1.5 for safety (i.e. absolute maximum value of 1.2 for the derailment coefficient).

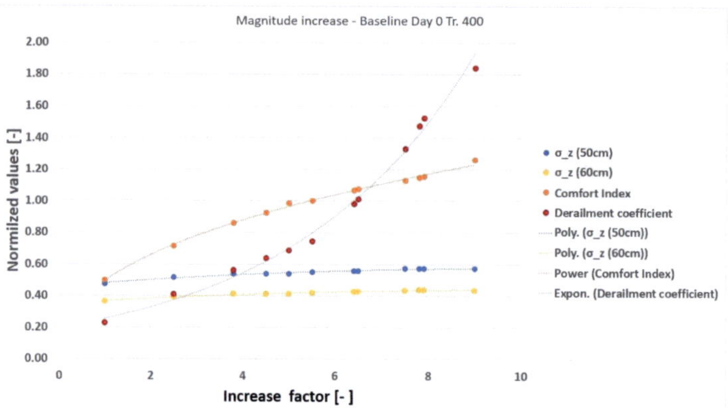

Figure 68:   **Magnitude increase based on track 400 to obtain refined limit values (own work)**

As it can be seen in the Figure 68, the comfort level dominates the limiting values up to an increase factor of 6.5. However, even when at this point the comfort level has reached a normalized value higher than 1, the safety factor has reached 1, as a consequence, safety is taken as the governing factor. Beyond this point, the derailment coefficient increases at a faster rate than the comfort level.

As seen in Figure 68, the absolute maximum value of 1.2 for the derailment coefficient was achieved at a magnitude of 7.8. This point represents a limit of immediate cease of operations for the target speed (speed restrictions/slow orders could be issued to decrease the risk of derailment). However, as it can be seen in the graph, the pressure exerted on the subsoil never reaches a critical level. Hence track damaging forces do not play an important role in limit determination.

In terms of operation of the track at 50 km/h, as it can be seen in Figure 69, the comfort limit does not reach the limit of very uncomfortable at magnitude increase F6.5 and F7.8; confirming that now safety takes precedence. Notwithstanding, it can also be seen that at 50 km/h the track had room to deteriorate further before an intervention was needed. This confirms the conclusion that the track might have been renewed early from the point of view of the comfort index (e.g. uncomfortable limit at comfort index 3.5) and safety (i.e. safety limit reached at around comfort index 3.76) for the vehicle under study.

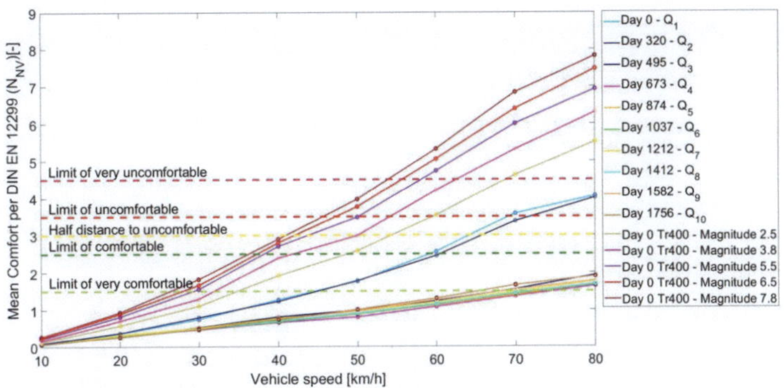

Figure 69:   **Comfort limit for "refined" magnitude increased signals (own work)**

Limits can be depicted in terms of PSD curves as well. Figure 70 shows a PSD plot comparing the PSD signals that achieved the expected comfort and safety limits for vertical irregularities against the PSD curves of the real signal for the worst TGQ, as well as the PSD curves from ERRI and FRA standards. The curves, for each parameter, were obtained from the average of 10 PSD curves and the convolution of the average per (Araji 2018).

As it can be seen, the PSD values are much higher than the worst day of the measured signals, which also confirms that there is still room for deterioration before the track is proposed for renewal. Also, it can clearly be seen that the FRA limits are well below the intervention limits determined, demonstrating that the conventional limits, even the worst quality (i.e. FRA class 1), are too stringent for the light rail system analyzed.

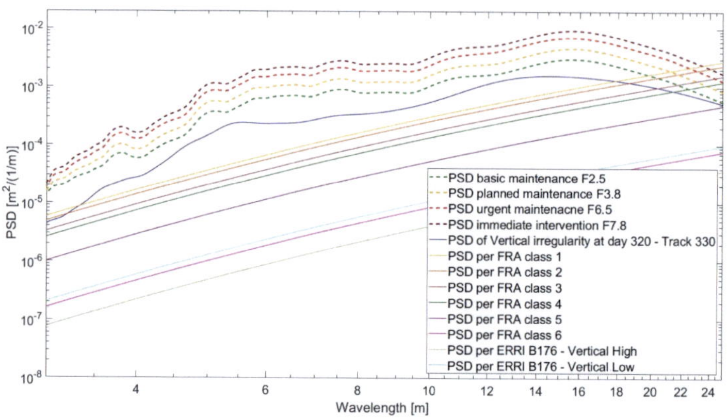

Figure 70: **PSD of limit values for vertical irregularities (own work)**

The limits determined can also be represented in terms of standard deviations. Table 27 shows the standard deviations limits for the vertical lateral irregularities.

Limits in this study	Vertical standard deviation developed	Horizontal standard deviation developed
Basic maintenance	13.6	1.7
Planned maintenance	16.8	2.1
Corrective Maintenance	21.9	2.7
Immediate action	24.0	3.0

Table 27: **Vertical and horizontal standard deviation limit values obtained in this study (own work)**

It is worth noticing the difference between the values obtained in this study and the values presented in standard EN 13848 – 5 (The British Standards Institution 2010), which only provides a standard deviation for an alert limit (AL). This AL can be compared to the point where the middle comfort is reached (i.e. where the track degrades from comfortable to middle comfort) or where the basic maintenance range ends. The comparison shows the largest difference for the vertical irregularities which according to Table 27, corresponds to 13.6 mm, while the value in the standards, according to Table 9, corresponds to maximum 3 mm. The difference is big (i.e. more than 4.5 times), providing an indication that the values in the standards are not adequate for the system studied.

A comparison between the values of horizontal irregularities shows that the horizontal irregularities are not as critical. The standard deviation obtained according to Table 27 is 1.7, while in EN 13848 – 5 (The British Standards Institution 2010), the alert limit value corresponds to 1.8 mm (see Table 9). This similarity is sensible from the point of view of the construction of the track, which is designed and built to maintain its horizontal alignment and to hold an appropriate gauge. Additionally, as it is known in this study, lateral irregularities do not deteriorate as intensively as vertical irregularities (e.g. lateral forces might not be high enough for a high / fast lateral deterioration of the track).

A useful standard deviation value corresponds to the magnitude increase of F5.5 (i.e. uncomfortable limit related); used as the urgent maintenance value in the calculation of the TGI (see section 3.4.3). It is worth noticing that although F5.5 was not used as a formal limit (i.e. only F2.5, F3.8, F6.5 and F7.8 were used), it was taken as the maximum value for the TGI's urgent maintenance since safety, represented by F6.5, should not be pushed to the limit.

The visualization options provided offer track/infrastructure managers choices to perform an evaluation of the condition of their asset, and hence to make decisions regarding its maintenance.

5.4 Track geometry index determination

The representations shown in the last section provide choices for evaluation to track managers, but their meaning might not be totally clear or simple to achieve. This is especially for the PSD graph, which might be mostly useful in specialized investigations. In addition, PSDs might even provide an unclear representation of the track against its limit values since PSD curves can span many orders of magnitude (e.g. PSDs from FFT) as seen in section 2.5.3.1 Figure 18.

Hence, as previously mentioned, the TGI might provide a practical and easy to read representation of the track condition and deterioration process. In this section, both tracks (i.e. track 330 and 400) are evaluated through the TGI method.

Since the history of track 330 is known, the TGI method should provide a clear picture of its condition at different points in time (see Figure 71). In case of track 400, which has mainly been used to determine limits of intervention, it will be used to provide a general idea of its deterioration process as well as a way to compare it to track 330. Lastly, both tracks' TGIs will be used to determine the point in time at which the track should be renewed through a deterioration graph (section 5.5).

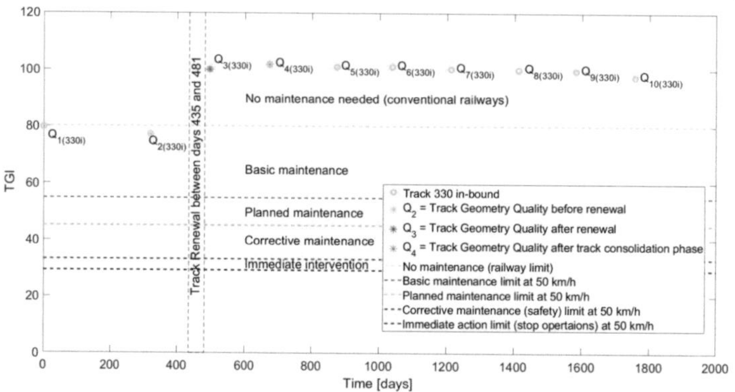

Figure 71:   **TGI representation for track 330 for a vehicle riding at 50 km/h (own work)**

The TGI performed on track 330 shows different phases of the track history. Points $Q_{1(330)}$ and $Q_{2(330)}$ show a low TGI (i.e. worst state of the track). The track renewal can be clearly seen after day 435 ($Q_{3(330)}$). Also, at point $Q_{4(330)}$ a further geometry improvement can be observed. This improvement was assumed to correspond to the construction method used by the transportation company, which performed a second tamping six week after the renewal on day 541(Camacho et al. 2016). This tamping is applied after a consolidation phase of the track, which occurs after 0.5-2 million tons of traffic (Lewis 2011). The deterioration process can be observed from $Q_{4(330)}$.

Interesting to point out in Figure 71, is that the track had not reached the "limit of comfortable" at its worst geometry. This means that the track before renewal was in relatively good condition with respect to the vehicle response at 50 km/h. Accordingly, if required, the track could have been renewed at a later time.

The limits in the graph show specific moments and actions required for intervention. The limit for no intervention was kept from regular railway lines which corresponds to a TGI of 80. The alert limit corresponds to the point at which the vehicle reaches the limit of comfort (i.e. comfort index of 2.5) for a magnitude increase of 2.5 and TGI of approximately 55. A warning limit was determined for the moment the vehicle reaction reaches half distance to the uncomfortable limit as determined from the comfort index (i.e. comfort index 3.0) reached at a magnitude increase of 3.8 or TGI 45. A corrective maintenance limit (safety related) was determined at a magnitude increase of 6.5 corresponding to a TGI of 33. Finally, a "stop operation" limit was achieved at magnitude increase of 7.8 or TGI 29. The following section will show the time between each limit within a 95% confidence level.

Track 400 shows a slightly different situation. As shown in Figure 72, the track experiences a slight improvement around TGQ $Q_{3(400o)}$, maintaining a similar quality level until TGQ $Q_{7(400o)}$. At this point, the track initiates its deterioration process and reaches, at $Q_{10(400o)}$, almost the same track quality as at TGQ observed for $Q_{1(400o)}$. No data were available to calculate the TGI for $Q_{8(400o)}$.

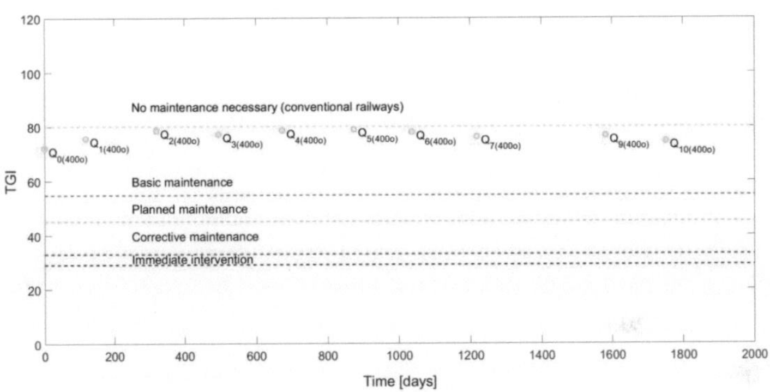

Figure 72: **TGI of track 400 (own work based on (Araji 2018))**

A comparison of TGIs between the two tracks shows mainly that track 400 displays in general a lower geometrical quality. For example, the TGI of track 400 is always below 80, while the TGI for track 330, after a renewal, lies always above a TGI of 80, even at its worst day after the renewal ($Q_{10(400o)}$).

The difference observed might be due to their corresponding track alignments. For example, for track 400 the presence of vertical and horizontal curves as well as slopes up to 6%. Another difference might be due to the construction tolerances used for the construction of the tracks, which might have been more stringent for track 330 than for track 400. Besides the fact that track segmentation was not performed for the analysis of track 400, a clue that tolerances might be laxer can be seen through the fact that track 400 is at all times below a TGI of 80, even after an apparent improvement of its geometry.

Logically, a factor determining the difference between the TGIs relates to the values chosen as the standard deviations for a new track. The values for this research were taken from the renewed track 330, which are lower than any value observed for track 400.

The reason for the use of higher tolerances for the construction of track 400 is not clear; however, it can be deduced that for track 400 to improve its TGI, the tolerance construction values should match the values of new construction implemented in track 330. Naturally, a segmentation and history of track 400 would be helpful to clarify the issues observed. This reinforces the idea that for an evaluation using real data, a complete record and segmentation of the track is necessary for the determination of the track's condition.

5.5   Track deterioration diagram

The track's deterioration rate can be determined from TGI values, but due to the low amount of available points to this date, a prognosis can only provide a limited accuracy. This is because as it can be seen in Figure 73, the 95% confidence level diverts greatly at about the sixth year. Moreover, a linear deterioration rate might not be completely accurate. Several studies show that the deterioration rate depicts rather a curved behavior. In this sense, it makes sense to keep monitoring the track measured data over time.

To have an idea of the future point of intervention of a track (maintenance or renewal), its deterioration is taken within the boundary between the linear deterioration line, and the lower boundary of the 95% confidence level (i.e. shaded area in Figure 73).

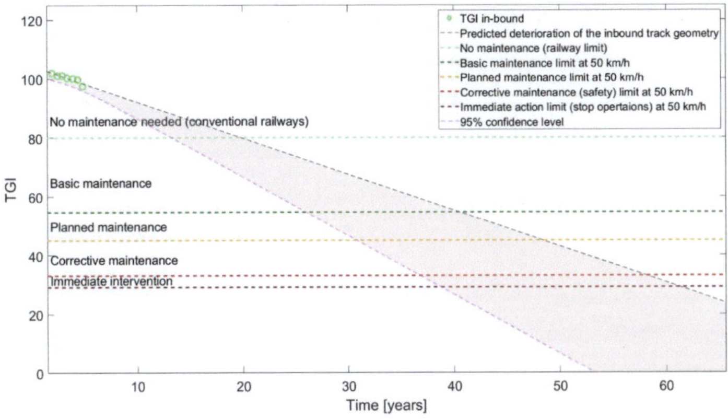

Figure 73:   **Deterioration process of track 330 (own work)**

As it can be seen, a short-term prognosis of one or two years ahead would be possible and would provide a good idea of the point in time that the track needs to be intervened to reestablish the required TGI. For longer periods of time, the range at which the track degrades widens, providing a wide range at which the track degrades. In this sense, the range of time at which the track reaches the limit of basic maintenance (i.e. entering a period of planned maintenance) is 26 to 40 years. On the other hand, the range of the time the track reaches the limit of planned maintenance (i.e. entering the period of corrective maintenance) is 31 to 48 years. Lastly, the range at which the track reaches the limit of corrective actions is between 37 to 58 years. After that, the track should be considered for immediate action (e.g. issuing slow orders or stop the operation), which is reached between 39 and 61 years. In general, the service life of a track is said to have a range of 30 to 50 years. Then, the time span obtained though the analysis is sensible, especially if it is considered that no infrastructure manager should let their track to reached dangerous limits (i.e. limits for corrective maintenance or immediate suspension of operations). In reality, infrastructure managers should look to reach maximum, and for short periods of time, TGIs between 33 (i.e. corrective maintenance) and 45 (i.e. planned maintenance). If a track is allowed to get to a higher degraded state, it would require performing corrective maintenance repetitively, which might not represent the best financial option. A track renewal would be then sensible.

For track 400 the prognosis over a long-time span is slightly more uncertain since the 95% confidence level tapered away faster from the linear fit. In addition, the fitted line is in general flatter, which projects the line to reach the limit of comfort between 15 and

29 years, the limit of planned maintenance between 21 and 40 years, a correction maintenance limit between 28 and 55 years and an immediate action limit between 30 and 59 years as seen in Figure 74.

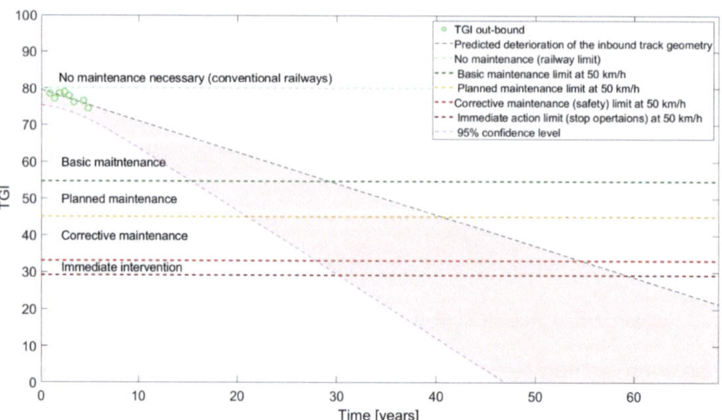

Figure 74:   Deterioration process of track 400 (own work per (Araji 2018))

The deterioration line of each track displays a different slope seen in Figure 73 and Figure 74. For track 330 the rate corresponds to a slope of -1.2243, while for track 400 the slopes correspond to -0.84924. These two figures show a comparison of the two tracks deterioration process from the highest TGI observed. The comparison is made in terms of the prognosis of track deterioration made above.

Furthermore, the tonnage experienced by track 330 is about 2.6 times higher than track 400. This is because track 330 operates three lines while track 400 operates only one.

However, although track 330 experiences more than twice load per year, it does not deteriorate twice as fast as track 400. This reinforces the idea that the initial quality of a track helps reduce the rate of deterioration. If track 400 was required to improve its service life and reduce its rate of deterioration, its initial quality after a renovation would have to be higher.

Track	Maintenance limit							
No.	Basic		Planned		Corrective		Immediate	
	Lower	Higher	Lower	Higher	Lower	Higher	Lower	Higher
	[years]	[years]	[years]	[years]	[years]	[years]	[years]	[years]
330	26	40	31	48	37	58	39	61
400	15	29	21	40	28	55	30	59

Table 28:   Track 330 and track 400 maintenance limits

5.6 Importance of limits and areas of opportunities

Limits of intervention based on available data are sensible for a more optimum and timely maintenance. The approach presented shows that track renewal activities are performed prematurely for the system under study. Premature renewal impacts the performance of the infrastructure and its overall costs. A timely intervention of the track (e.g. knowledge-condition-based maintenance), might help reducing the LCC of infrastructure. This in turn helps promoting the system as a viable solution; especially in countries that use the high cost of investment as an argument against LRT systems.

To learn about the actual condition of the track, it is important to gather track geometry and vehicle reaction data and properly process them to obtain TGI and PSD plots which graphically show the deterioration and degradation of the track. Data should be obtained for a longer period of time and ideally should be done or performed on in-service vehicles which allows for a more reliable prediction of track deterioration and degradation, or even serve as a calibration of previously made predictions. However, as instrumentation of in-service vehicles might represent a challenge. Hence, working with track recording vehicles might be the first step towards achieving a systematic approach to maintenance optimization. In this sense, measurements of track geometry should be conducted at least twice a year so that the amount of data is large enough to allow for better predictions. This in turn would allow infrastructure managers to develop better maintenance strategies and regimes.

If no instrumentation on in-service vehicles is performed, to be able to establish the appropriate limits, an MBS model of the vehicle should be done to obtain its reactions to the measured track geometry. By now, it should be obvious that expertise in signal processing is essential to deal with the data as well as to interpret them to determine the limits and to plot them on the TGI or PSD graph against continuously gathered data. To avoid a possible issue of expertise to process and interpret the required data, it would be recommended that measured track geometry signals and vehicle accelerations are preprocessed accordingly within the track recording vehicle / or in-service vehicle through algorithms within an application that allows the results to be appropriately handled and presented. The idea is that an infrastructure manager obtains the results as easier to read and interpret as possible (as seen in Figure 71 to Figure 74).

6 Conclusions and Future Work

The notion that LRT systems require specific limits for maintenance, inherently different from regular railways, has been positively determined and shown in this work. However, due to aspects that govern different urban systems (e.g. budget), the limits should not be standard, but rather determined for each system in consideration to the specific characteristics and needs of each system, i.e. vehicles operated, maximum and operating speeds, etc. Experienced based approaches to determine maintenance are without a doubt helpful to determine a somehow appropriate time to perform the maintenance or renewal of a track. However, it is not appropriate to generate an optimum maintenance strategy. This is especially true when technological advancements cause systems to behave differently than in the past (e.g. less stiff suspension or infrastructure). Track condition-based knowledge might help develop more optimal maintenance plans, especially if track irregularity data is available for several years.

Measuring track data through track recording vehicles is a modern practice that provides the bases of knowledge-based maintenance and renewal. Data can be gathered through track recording vehicles using different measuring systems, namely inertial and chord systems. The treatment of the data is important if results are expected to be reliable. The data treatment is based on signal processing methods in both the spatial and frequency domain. Based on this discussion, if an attempt to optimize the maintenance processes is pursued, it is important to implement knowledge-based approaches.

The track recording vehicle in Stuttgart, Germany, is able to provide information of all track parameters in spatial and frequency domain. The frequency domain is currently not done, but possible on demand which would facilitate the implementation of the method developed in this work. However, several aspects need to be considered regarding the data treatment. First, they need to have a homogeneous sampling rate for the analysis as well as the domain of the data (i.e. time / spatial or frequency). The track recording vehicle company could directly provide frequency domain data and apply the required filters (i.e. filter the D1 wavelengths). Data from inertial and chord systems would be beneficial to compare results, but the chord data needs to be decolored to eliminate the influence of the measuring chord's geometry.

In addition, an improvement to the cross-level irregularity would be beneficial. The track recording company could provide more accurate short wavelength information. Currently, the data for wavelengths lower than three meters show unusually high PSD magnitudes which cannot be confirmed on the real track. This issue needs to be addressed in the future. The unusually high irregularities might be the result of the track recording used. A basic idea is to place the measuring devices on a train similar to the ones in operation, which possess similar characteristics in terms of speed, total weight, eigenfrequencies of the suspension, vibrating behavior, etc. Another idea could be to measure the track with handheld devices which concentrate their measurements in short wavelengths.

In fact, monitoring the track continuously would also be beneficial. This way effects of maintenance and other parameters, such as temperature differences, could be considered. Continuous measurements would also confirm the deterioration trend or adjust it. This was discussed in chapter 5 where it was determined that the deterioration diagram is able to provide a fairly good prediction some 2 – 3 years in advance. Having a continuous measurement (i.e. on in-service vehicles) and a large number of years would increase the certainty of the prediction providing a relatively good time for infrastructure managers to react to track problems or to plan for maintenance activities and budgets.

This work concentrated in the condition of the track, assuming a static condition of the vehicle (called here nominal condition). However, the vehicle deterioration should be also researched along with the track. Several approaches can be used to attempt it. In fact, in future studies at the University of Stuttgart, fault tolerant control and fault detection and isolation algorithms would be implemented to detect problems in the track and vehicles. Through these methods, patterns of behavior will be established, forming the basis of a behavior catalog which could help detect problems on in-service trains through machine learning algorithms. Hence, not only condition and prediction-based maintenance for tracks would be possible, but also maintenance programs for vehicles could be developed.

To achieve good results in the above-mentioned research, instrumentation of in-service vehicles will be carried out. These efforts will be part of the following steps to be followed in the near future in relation to ongoing research efforts.

Glossary

Angle of Attack (α) The angle of attack α is at the horizontal sectional plane through the wheel flange / rail contact point. It is the angle between the tangent at the rail in the contact point and the right angle to the axle (Office of Rail and Road 2004)

Amplitude Half peak to peak of a signal (Rossiter 2013)

Axle Load Total weight felt by the roadway for all wheels connected to a given axle

Capital Costs The capital costs comprise the consumption of fixed capital (depreciation costs) and interest. Capital costs represent a high share of total infrastructure costs and are different to the annual capital expenditures (Transport and Delft 2005)

Degrees of Freedom Minimum number of variables necessary to fully describe the configuration of a system (Flores 2015)

Ergodic The mean and variance of a signal can be derived from any sufficiently long sample of the signal

Frequency Response Description of how the amplitude and phase shift of the output, relative to the input, change with frequency (Rossiter 2013)

Gain Ratio of the output amplitude to the input amplitude (Rossiter 2013)

Initial Settlement Rapid loss of track quality within the first MGT of traffic (Lichtberger 2001)

Power Spectral Density The magnitude squared of the Fourier Transform of a continuous time and finite power signal. It is the quantity of power for each frequency component: therefore, PSD integral (in frequency domain) is the total signal power. In simple terms, it describes the frequency-dependent power content of a stationary signal. Power Spectral Density is a real function of the frequency but does not contain phase information.

	The word power in this case means the square of the amplitude of the signal. Density means the actual power is not returned but rather a power per frequency range. The total power can be determined by integrating the PSD over the frequency (Dassault Systems Simulia Corp. 2017)
Signal	Entire data recording
Spatial Domain Signals	Physical signals or spatial series of data, with respect to spatial dimensions (e.g. millimeters)
Track Geometry Quality	The deviation of the spatial position of the rails against their theoretical position (Luber et al. 2010).
Track Irregularity	The deviation of the spatial position of a parameter describing the track geometry against it designed position
Track Maintenance	The product of resources, judgments, experience, skills, tools, and policies that are exercised in a range of service environments and within every conceivable type of organizational structure (Daniels 2008).
Track Structural Quality	The deviation from the acceptable mechanical behavior of the track structural components.
Wavelength	Spatial period of a periodic wave which represents the distance over which the wave's shape repeats
Wavenumber	The spatial frequency of a wave, either in cycles per unit distance or radians per unit distance (also named angular wavenumber or circular wavenumber). It is analogous to the temporal frequency being the number of cycles or radians per unit time. The International Standard Organisation (ISO) recommends the use of repetency instead of wavenumber.
	Per ISO 80000:3: In English, the names repetency and angular repetency should be used instead of wavenumber and angular wavenumber, respectively, since these quantities are not numbers.

References

1. Verband Deutscher Verkehrsunternehmen (2014) Stadtbahnsysteme/Light Rail Systems: Grundlagen - Technik - Betrieb - Finanzierung/Principles - Technolo-gy - Betrieb - Finanzierung, 1. Aufl. DVV Media Group -Eurailpress, Hamburg

2. Parsons B (2012) Track Design Handbook for Light Rail Transit, Second Edi-tion, Washington, D.C.

3. Gordon P, Richardson HW, Holcombe R et al. (2001) Transportation and land use. Smarter Growth: Market-Based Strategies for Land Use Planning in the 21st Century: 27–57

4. Kochs A, Marx A (2009) Innovatives Instandhaltungsnabagement mit IDMVU: Teil 1 - Überblick Gesamtprozess. Infrastrukture-Daten-Management für Verkehrsunternehmen, Version 1.0

5. Daniels LE (2008) Track maintenance costs on rail transit properties

6. Tzanakakis K (2013) The railway track and its long term behaviour: A hand-book for a railway track of high quality. Springer Tracts on Transportation and Traffic, Vol. 2. Springer Science & Business Media, Heidelberg

7. Deutsches Institut für Normung e. V. (2009) Railway applications - Ride com-fort for passangers – Measurment and evaluation; BS EN 12299:2009 (DIN EN 12299:2009), Beuth Verlag

8. Ferreira L, Murray MH (1997) Modelling rail track deterioration and mainte-nance: current practices and future needs. Transport Reviews 17(3): 207–221

9. Xu P, Liu R-K, Wang F et al. (2013) Railroad track deterioration characteristics based track measurement data mining. Mathematical Problems in Engineering 2013

10. International Organization for Standardization, ISO 9000 (2015) Quality Management Systems - Fundamentals and Vocabulary. International Organization for Standardization

11. Deutsches Institut für Normung e. V. (2016) Railway applications - Track - Track geometry quality - Part 1: Characterisation of track geometry; German and English version prEN 13848-1:2016 (DIN EN 13848-1:2016), Beuth Verlag

12. Suiker ASJ (2004) The mechanical behaviour of ballasted railway tracks

13. Camacho D, Le TH, Rapp S et al. (2016) Light rail ballasted track geometry quality evaluation using track recording car data. Computers in Railways XV: Railway Engineering Design and Operation 162: 303

14. Sadeghi J, Askarian AH(2007) Influences of track structure, geometry and traffic parameters on railway deterioration

15. Haigermoser A, Eickhoff B, Thomas D et al. (2014) Describing and assessing track geometry quality. Vehicle System Dynamics 52(sup1): 189–206. doi: 10.1080/00423114.2014.889318

16. Berawi ARB, Delgado R, Calçada R et al. (2010) Evaluating Track Geomet-rical Quality through Different Methodologies 1(1). http://www.ijtech.eng.ui.ac.id/index.php/journal/article/view/35

17. Veit P (2007) Track quality-luxury or necessity? Railway technical review/RTR special (July): 8–12

18. Iyengar RN, Jaiswal OR (1995) Random field modeling of railway track irregularities. Journal of Transportation Engineering 121(4)

19. Garg VK, Dukkipati RV (1984) Dynamics of railway vehicle systems

20. Knothe K, Grassie SL (1993) Modelling of Railway Track and Vehicle/Track Interaction at High Frequencies. Vehicle System Dynamics 22(3-4)

21. Karis T (2009) Track irregularities for high-speed trains: Evaluation of their correlation with vehicle response. Master of Science Thesis, Royal Institute of Technology

22. Haigermoser A (2013) Dyno TRAIN - Railway Vehicle Dynamics and Track Interactions: Total Regulatory Acceptance for Interoperablre Nwetwork: D2.6 - Final Report on Track Geometry. WP2 - Track Geometry Quality

23. Li MXD, Berggren EG, Berg M et al. (2008) Assessing track geometry quality based on wavelength spectra and track–vehicle dynamic interaction. Vehicle Sys-tem Dynamics 46(S1): 261–276

24. Lewis R (2011) Track geometry recording and usage. Notes for a lecture to Network Rail

25. Deutsches Institut für Normung e. V. (2008) Railway applications - Track - Track geometry quality - Part 1: Characterisation of track geometry; German version EN 13848-1:2003+A1:2008 (DIN EN 13848-1:2008-11 (E)) Beuth Verlag

26. Deutsches Institut für Normung e. V. (2010) Railway applications - Track – Track geometry quality – Part 5 Geometric quality levels – Plain line; German version EN 13848-5 2008+A1 2010 (EN 13848-5 2008+A1 2010) Beuth Verlag

27. The British Standards Institution (2014) Railway applications - Track - Track geometry quality - Part 6: Characterisation of track geometry (BS EN 13848-6:2014)

28. The British Standards Institution (2016) Railway applications - Testing and Simulation for the acceptance of running characteristics of railway vehicles - Running Behaviour and stationary tests (BS EN 14363:2016)

29. DB Netz AG: Bautechnik, Leit-, Signal-, u. Telekommunikationstechnik (2004) Oberbau inspizieren: Prüfung der Gleisgeometrie mit Gleismessfahrzeugen (Ril 821.2001)

30. The British Standards Institution (2009) Railway applications - Ride comfort for passengers - Measurement and evaluation; English version BS EN 12299:2009 (BS EN 12299:2009)

31. The British Standards Institution (2010) Railway applications - Track - Track geometry quality - Part 5: Geometric quality levels - Plain line (BS EN 13848-5:2008+A1:2010)

32. Pollard MG, Simons NJA (1984) Passenger comfort—the role of active suspensions. Proceedings of the Institution of Mechanical Engineers, Part D: Transport Engineering 198 (3): 161–175

33. International Organization for Standardization (1997) Mechanical vibration and shock: Evaluation of human exposure to whole-body vibration. Part 1: General requirements. ISO 2631-1:1997 (E). International Organization for Standardization

34. Dassault Systems Simulia Corp. (2017) SIMPACK Documentation: D.19 Power Spectral Densities. Description, SIMPACK Release 2017.2, vol 6

35. Benz S (2016) Klassifizierung auffälliger Messwerte anhand des HuDe - Gleis-messsystems der SSB AG. Master's, University of Stuttgart

36. Fazio AE, Corbin JL (1986) Track quality index for high speed track. Journal of Transportation Engineering 112 (1): 46–61

37. Sadeghi J (2010) Development of railway track geometry indexes based on statistical distribution of geometry data. Journal of Transportation Engineering 136 (8): 693–700

38. Zhang Y-J, El-Sibaie M, Lee S (2004) FRA track quality indices and distribu-tion characteristics

39. Lee S (2005) Development of objective track quality indices. Federal Railroad Administration, Research Results, RR: 05-01

40. Liu R-K, Xu P, Sun Z-Z et al. (2015) Establishment of Track Quality Index Standard Recommendations for Beijing Metro. Discrete Dynamics in Nature and Society

41. Research Design and Standards Organisation (1997) Implementation of Track Geometry Index (TGI) - Annexure-F-2 (No. 94/Track-III/TK/23). http://122.252.243.98/Departments/openline/cnm/Manuals/Enginfo/mauals/ESO/Eso_6/Annexure/Annex_F2.html. Accessed 14 May 2018

42. Talukdar KN, Arulmozhi U, Prrabhakar K et al. (2006) Project presentation: Improving of TGI by using computer analysis

43. Sadeghi J, Askarinejad H (2010) Development of improved railway track degradation models. Structure and infrastructure engineering 6 (6): 675–688

44. Chudzikiewicz A, Bogacz R, Kostrzewski M et al. (2017) Condition monitoring of railway track systems by using acceleration signals on wheelset axle-boxes. Transport: 1–12

45. Hyslip J (2002) Fractal analysis of track geometry data. Transportation Re-search Record: Journal of the Transportation Research Board (1785): 50–57

46. Luber B (2011) Methoden zur Bewertung von Gleislageabweichungen auf Basis von Fahrzeugreaktionen. Dissertation - Doctoral, Technischen Universität Graz

47. Luber B, Haigermoser A, Grabner G (2010) Track geometry evaluation method based on vehicle response prediction. Vehicle System Dynamics 48 (sup1): 157–173. doi: 10.1080/00423111003692914

48. Li D, Meddah A, Hass K et al. (2006) Relating track geometry to vehicle performance using neural network approach. Proceedings of the Institution of Mechan-ical Engineers, Part F: Journal of Rail and Rapid Transit 220 (3): 273–281

49. Stoica P, Moses RL (2005) Spectral analysis of signals. Pearson/Prentice Hall, Upper Saddle River, N.J.

50. Harris FJ (1978) On the use of windows for harmonic analysis with the dis-crete Fourier transform. Proc. IEEE 66(1): 51–83. doi: 10.1109/PROC.1978.10837

51. Cimbala JM (2007) Windowing with FFTs. ME 345 Instrumentation, Meas-urements and Statistics

52. Barry Van Veen (2013) The Power Spectral Density, All Signal Processing - Online lecture

53. Mathworks (2018) Signal Processing Toolbox™. User's Guide (R2018a). www.mathworks.com/help/pdf_doc/signal/signal_tb.pdf. Accessed 31 May 2018

54. Berawi ARB (2003) Improving Railway Track Maintenace Using Power Spec-tral Density (PSD). Doctor of Philosophy in Transport Systems, Universidade do Porto

55. Liu D (2015) The influence of track quality to the performance of vehicle track interaction, TUM, Technische Universität München, Lehrstuhl und Prüfamt für Verkehrswegebau

56. Lei X (2017) Track Irregularity Power Spectrum and Numerical Simulation. In: High Speed Railway Track Dynamics: Models, Algorithms and Applications. Spring-er Singapore, Singapore, pp 137–160

57. Ilvedson CR (1998) Transfer function estimation using time-frequency analy-sis. M. Sc., Massachusetts Institute of Technology

58. Jiang X, Luo Z-Q, Georgiou TT (2012) Geometric methods for spectral analy-sis. IEEE Transactions on Signal Processing 60(3): 1064–1074

59. Di Scalea FL, McNamara J (2004) Measuring high-frequency wave propaga-tion in railroad tracks by joint time–frequency analysis. Journal of sound and vibra-tion 273(3): 637–651

60. Mallat S (1999) A wavelet tour of signal processing. Academic press

61. Zhiping Z, Fei L, Yong Z (2010) Wavelet analysis of track profile irregularity for Beijing-Tianjin intercity high speed railway on bridge: 1155–1158

62. The British Standards Institution (2006) Railway applications - Track - Track ge-ometry quality - Part 2: Measuring Systems - Track recording vehicles (BS EN 13848-2:2006)

63. HuDe Mess- & Anlagentechnik GmbH (2011) Handbuch Gleismesssystem: Be-nutzerhandbuch

64. Mathworks (2018) Documentation: fillgaps. www.mathworks.com/help/sig-nal/ref/fillgaps.html

65. Mathworks (2018) MATLAB® Data Analysis. www.math-works.com/help/pdf_doc/matlab/data_analysis.pdf

66. Podwórna M (2015) Modelling of Random Vertical Irregularities Of Railway Tracks. International Journal of Applied Mechanics and Engineering 20(3): 87. doi: 10.1515/ijame-2015-0043

67. Dumitriu M (2014) Method to synthesize the track vertical irregularities. Scien-tific Bulletin of the" Petru Maior" University of Targu Mures 11(2): 17

68. Dumitriu M (2016) Numerical synthesis of the track alignment and applica-tions. Part I: The synthesis method. Transport Problems 11(1): 19–28

69. Lestoille N, Clouteau D, Degrande G et al. (2015) Stochastic model of high-speed train dynamics for the prediction of long-time evolution of the track irregulari-ties. [s.n.], [S.l.]

70. Nico Wächtler Assesment of Stochastic Track Irregularities from Tracks. SIM-PACK News (August 2014): pp 10 -11

71. Perrin G, Soize C, Duhamel D et al. (2013) Track irregularities stochastic modeling. Probabilistic Engineering Mechanics 34: 123–130

72. Salcher P (2015) Reliability assessment of railway bridges designed for high-speed traffic: Modeling strategies and stochastic simulation. Innsbruck, Univ., Diss., 2015

73. Araji O (2018) Determination of suitable combinations of artificially generated track geometry for the determination of limits for the maintenance and renewal of light rail tracks. Master's Thesis, University of Stuttgart

74. Álvarez A (2018) Characterization of Light Rail Tracks through Fractal and Frequency Domain Methods. Master's Thesis, University of Stuttgart

75. Esveld C Track structures in an urban environment

76. Verband Öffentlicher Verkehrsbetriebe (2008) German Federal Regulations on the construction and operation of light rail transit systems: Straßenbahn-Bau- und Betriebsordnung - BOStrab issued 11th December 1987 (English Translation). Einkaufs- und Wirtschaftsges. für Verkehrsbetriebe

77. Indraratna B, Rujikiatkamjorn C, Vinod JSJ et al. (2009) A review of ballast characteristics, geosynthetics, confining pressures and native vegetation in rail track stabilisation

78. Kerr AD (2002) The determination of the track modulus k for the standard track analysis

79. Kennedy J (2011) A full-scale laboratory investigation into railway track sub-structure performance and ballast reinforcement, Heriot-Watt University

80. Hesse DE, Tinjum JM, Warren BJ (2014) Impact of increasing freight loads on rail substructure from fracking sand transport. Transportation Geotechnics 1(4): 241–256

81. Li D, Hyslip J, Sussmann T et al. (2016) Railway geotechnics. CRC Press / Taylor and Francis Group, Boca Raton, FL

82. Martin U, Rapp S, Camacho D et al. (2016) Abschätzung der Untergrundverhältnisse am Bahnkörper anhand des Bettungsmoduls

83. Powrie W, Louis LP (2016) A Guide to Track Stiffness, Southhampton, UK

84. Göbel C, Lieberenz K (2004) Handbuch Erdbauwerke der Bahnen:[eisenbahn-technische und geotechnische Grundlagen; Planung, Bemessung und Ausführung; Schutzschichten, Entwässerung, Ertüchtigung, Instandhaltung]. Eurailpress Tetzlaff-Hestra

85. Rapp S (2017) Modell zur Identifizierung von punktuellen Instabilitäten am Bahnkörper in konventioneller Schotterbauweise. Dissertation, Universität Stuttgart

86. Esveld C Modern railway track. MRT-Productions, Duisburg, Germany. MRT - Productions, Duisbug

87. Praticò FG, Giunta M (eds) (2016) Assessing the sustainability of design and maintenance strategies for rail track by means life cycle cost analysis

88. Praticò FG, Giunta M (2017) An Integrative Approach RAMS-LCC to Support Decision on De-sign and Maintenance of Rail Track

89. Kaewunruen S, Remennikov A (2008) Dynamic properties of railway track and its components: A state-of-the-art review

90. Wenty R (2007) Latest Developments in Track Rehabilitation and Mainte-nance: Maintenance and Renewal RTR - Special. Journal Railway Technical Re-view 47 (July 2007): 37–42

91. Lichtberger BW (2007) Railway track optimisation by efficient track mainte-nance machinery and strategies. Rail Engineering International 36 (4)

92. Lichtberger B (2005) Track Compendium, 2nd. Tetzlaff-Hestra GmbH & Co. Publ, Hamburg

93. Lichtberger B (2007) The Track System and Its Maintenance: Maintenance and Renewal RTR - Special. Journal Railway Technical Review 47 (July 2007): 14–22

94. Lichtberger B (2005) Comparison LCC Cuts Track Maintenance Cost. International Railway Journal 45 (9)

95. Lichtberger H (2001) Track maintenance strategies for ballasted track-a selection. Rail Engineering International 30 (2)

96. Schilling R (2005) Ballast cleaning of single-track railway lines: A strategic analysis. Rail Engineering International 34 (1)

97. Popovic Z, Lazarevic L, Vukiceivc M et al. (2017) The modal shift to sustaina-ble railway transport in Serbia

98. Flores P (2015) Concepts and formulations for spatial multibody dynamics. Springer

99. Schupp G, Weidemann C, Mauer L (2004) Modelling the Contact Between Wheel and Rail Within Multibody System Simulation. Vehicle System Dynamics 41 (5): 349–364. doi: 10.1080/00423110412331300326

100. Holzscheiter M (2017) Ermittlung von Toleranzwerten für die Gleisinstandhaltung in Straßen- und Stadtbahnnetzen mittels Mehrkörpersimulation. Studienarbeit

101. Rill G, Schaeffer T (2010) Grundlagen und Methodik der Mehrkörpersimulation. Springer

102. Bezin Y, Iwnicki SD, Cavalletti M et al. (2009) An investigation of sleeper voids using a flexible track model integrated with railway multi-body dynamics. Proceedings of the Institution of Mechanical Engineers, Part F: Journal of Rail and Rapid Transit 223(6): 597–607. doi: 10.1243/09544097JRRT276

103. Kalker JJ (1982) A Fast Algorithm for the Simplified Theory of Rolling Contact. Vehicle System Dynamics 11 (1): 1–13. doi: 10.1080/00423118208968684

104. Vogel W, Lieberenz K, Neidhart T et al. (2015) Eisenbahnstrecken mit Schotteroberbau auf Weichschichten-Rechnerisches Verfahren zur Untersuchung der dynamischen Stabilität des Eisenbahnfahrwegs bei Zugüberfahrten

105. Skorsetz S, Strobel T, Camacho Alcocer D et al. (September 23 -27) Influence of corrugation on light rail vehicles in terms of ride quality and safety

106. Strobel T, Skorsetz S, Camacho Alcocer D et al. Methode zur Bewwertung der Ineraktion Stadtbahnfahrzeug / Infrastruktur - Wie weit lassen sich Grenzwerte ausreizen?: Method for Assessing the Vehicle / Infrastructure Interaction in Light-Rail Systems - How much can Existing Limit Values of Intervention be Exceeded?, vol 141. Georg Siemens Verlag, Berlin

107. Frohling RD, Scheffel H, Ebersöhn W (1996) The Vertical Dynamic Response of a Rail Vehicle caused by Track Stiffness Variations along the Track. Vehicle System Dynamics 25 (sup1): 175–187. doi: 10.1080/00423119608969194

108. (2004) Desing Principles Applicable to the Guidance of Rail Vehicles for use on UK Tramways - Regulations on the Guidance of Rail Vehicles in accordance with the German Federal Regulations on the Construction and Operation of Light Rail Transit Systems (BOStrab): SpR

109. Rossiter J (2013) Bode diagrams 1 - basic concepts and illustration of fre-quency response. Bode diagrams, YouTube

110. Transport E, Delft CE (2005) Infrastructure expenditures and costs. Practical guidelines to calculate total infrastructure costs for five modes of transport. Final re-port. Retrieved from: http://ec. europa. eu/transport/themes/infrastructure/stud-ies/doc/2005_11_30_guidelines_infrastructu re_report_en. pdf

111. Selig ET, Waters JM (1994) Track geotechnology and substructure manage-ment. T. Telford; American Society of Civil Engineers, Publications Sales Dept. [dis-tributor], London, New York

112. Berggren E (April /2009) Railway Track Stiffness: Dynamic Measurements and Evaluation for Efficient Maintenance. Doctoral Thesis, KTH Royal Institute of Technology

113. Giannakos K (2010) Loads on track, ballast fouling, and life cycle under dy-namic loading in railways. Journal of Transportation Engineering 136 (12): 1075–1084

114. Bauer G, Theurer U (2000) Von der Straßenbahn zur Stadtbahn Stuttgart 1975 - 2000: Eine Dokumentation über den Schienenverkehr der Stuttgarter Straßen-bahnen AG (SSB) zwischen 1975 und 2000, Stuttgarter Straßenbahnen Band IV. Franz Kaufmann GmbH, SSB, Stuttgart

115. Punmia BC, Jain AK (2005) Soil Mechanics and Foundations. Laxmi Publica-tions Pvt Limited

116. Schmutz G (1980) Ein Beitrag zur Dimensionierung des Eisenbahnunterbaus. Schweizer Ingenieur und Architekt 30: 31

Abbreviations

AG	Aktiengesellschaft (Eng. Joint-Stock Company)
AI	Alignement Index
AL	Alert Limit
AMTRAK	National Railroad Passenger Corporation (USA)
CL	Cross Level (Track parameter)
CM	Corrective Maintenance
COM	Center of Mass
CTR	Composite Track Record
CWR	Continuous Weld Rail
DB AG	Deutsche Bahn AG
DFT	Discrete Fourier Transform
DOF	Degrees of Freedom
Eng.	English
ERRI	European Rail Research Institute
FFT	Fast Fourier Transform
FRA	Federal Railway Administration
Ger.	German
GI	Gauge Index
GPS	Global Positioning System
IAL	Immediate Action Limit
.if2	ASCII files used as input functions in SIMPACK©
IL	Intervention Limit
LCC	Life Cycle Cost
LRT	Light Rail Transit
LRV	Light Rail Vehicle
KFT	Kombinierte Frostschutz – Tragschicht
MBS	Multibody Simulation
MDZ	mechanisierter Durcharbeitungszüge (Eng. Mechanized Processing Trains)
MGT	Million Gross Tons

MIMO	Multiple-input and multiple-output
NaN	Not a Number
n.d.	Not described
ÖBB	Austrian Federal Railways
PBTG	Performance-Based Track Geometry
PDM	Predictive Maintenance
PM	Preventive Maintenance
PSD	Power Spectral Density
RAMS	Reliability, Availability, Maintainability and Safety
RFID	Radio Frequency Identification
RMS	Root Mean Square
RTF	Run to Failure
SD	Standard Deviation
SSB	Stuttgarter Straßenbahnen AG
SR	Störgröße / Reaktion (Eng. Disturbance / Reaction)
STFT	Short-Time Fourier Transform
TF	Transfer Function
TGA	Track Geometry Assessment
TGI	Track Geometry Index
TGQ	Track Geometry Quality
TI	Twist Index
TQ	Track Quality
TTF	Typical Transfer Function
UI	Unevenness Index (Vertical Irregularity Index)
VRA	Vehicle Response Analysis
WGB	wahrgenommene Gesundheitsbedrohung (Eng. Perceived Health Threat)
WT	Wavelet Transformation

Symbols

A_s	Contact area of sleeper minus pressure free zone in the middle in [m^2]
b_A	Lateral spacing of rail-wheel contact points in [m]
C	Total Track modulus in [MN/m^3]
C_g	Track modulus of the ground in [MN/m^3] (C_b in (Rapp 2017))
C_{rp}	Track modulus of the rail pad in [MN/m^3] (C_{zw} in (Rapp 2017))
δ	Deflection [mm]
f_{veh}	Eigenfrequency of vehicle in [Hz]
k_{rp}	Rail pad stiffness in [MN/m]
λ	Wavelength of track irregularities in [m]
ω	Angular wavenumber in spatial domain in [2π/m] (Frequency in time
ρ	Density of the soil [g/mm^3]
ρ_w	Wet density of the soil [g/mm^3]
q	Force per unit length [N/m]
$S(\omega)$	Power Spectral Density in [mm^2 /(2π/m)]
s	Sagitta or versine
u	Track modulus [MPa]
υ	Poisson ratio [-]
v_{veh}	Speed of the vehicle in [m/s]
V_{veh}	Speed of the vehicle in [km/h]
$\tilde{\upsilon}$	Wavenumber in [1/m]
N_{MV}	Mean Comfort Standard Method in [m/s^2]
N_{VD}	Mean Comfort Complete Method in [m/s^2]
$\ddot{y}_{axle}$	Lateral axle box acceleration in [m/s^2]
$\ddot{y}_{cabin}$	Lateral passenger cabin acceleration in [m/s^2]
$\ddot{z}_{axle}$	Vertical axle box acceleration in [m/s^2]
$\ddot{z}_{cabin}$	Vertical passenger cabin acceleration in [m/s^2]

Appendix I: Track Irregularities

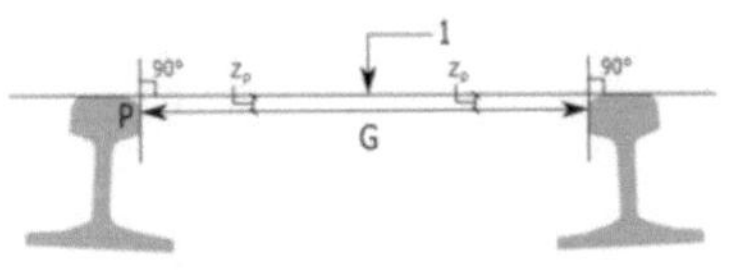

Gauge (G)

The smallest distance between lines perpendicular to the running surface intersecting each rail head at point P in a range from 0 to a fixed distance Z_p (14 mm) below the running surface (1)

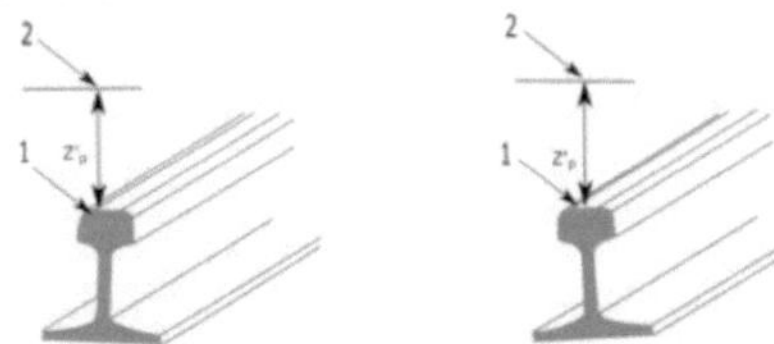

Vertical Profile (Longitudinal level)

Is the deviation $Z_p{'}$, of consecutive running table (1) levels on any rail (i.e. left or right rail), expressed as an excursion from the mean vertical position of a reference line (2)

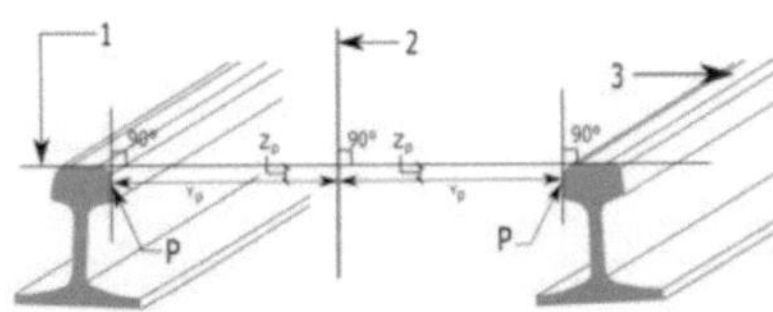

Lateral Alignment

Is the deviation Yp in y-direction of consecutive positions of point P on any rail, expressed as an excursion from the mean horizontal position from a reference line (2) and the center line of a track

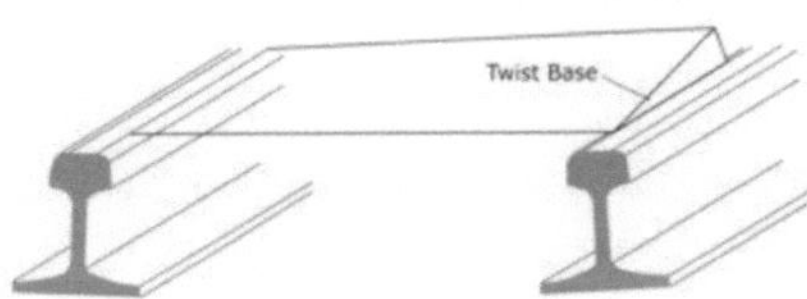

Twist

Twist is the algebraic difference between two cross levels taken at a defined distance apart.

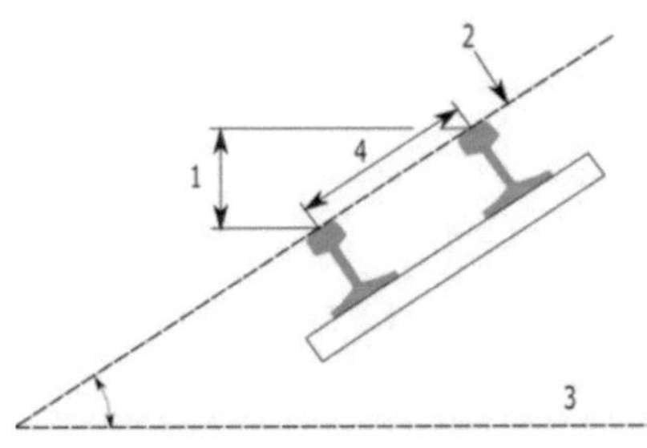

Cross-level

Is the difference in height (1) of adjacent running tables computed from the angle between the running surface (2) and a horizontal reference plane (3). It is expressed as the height of the vertical leg of the right-angled triangle having a hypothenuse (4) that relates to the nominal track gauge plus the with of the rail head

Figure 75:   **Track Geometry Parameters (own work per (Deutsches Institut für Normung e. V. 2016))**

Appendix II: Conversion between Coordinate Systems

Appendix II.A: Conversion from rail to track coordinate systems

$$z_{Cl}(x) = \frac{z_L(x) + z_R(x)}{2} \tag{31}$$

$$y_{CL}(x) = \frac{y_L(x) + y_R(x)}{2} \tag{32}$$

$$\delta(x) = \frac{z_R(x) - z_L(x)}{2b_A} \tag{33}$$

$$g(x) = g_o + y_R(x) - y_L(x) \tag{34}$$

Appendix II.B: Conversion from track to rail coordinate systems

$$z_L(x) = z(x) - b_A\delta(x) \tag{35}$$

$$z_R(x) = z(x) + b_A\delta(x) \tag{36}$$

$$y_L(x) = y(x) + \frac{g_o - g(x)}{2} \tag{37}$$

$$y_R(x) = y(x) - \frac{g_o - g(x)}{2} \tag{38}$$

Appendix III: Mean Comfort Calculation

The process to calculate the mean comfort index N_{MV} is shown in Figure 76:

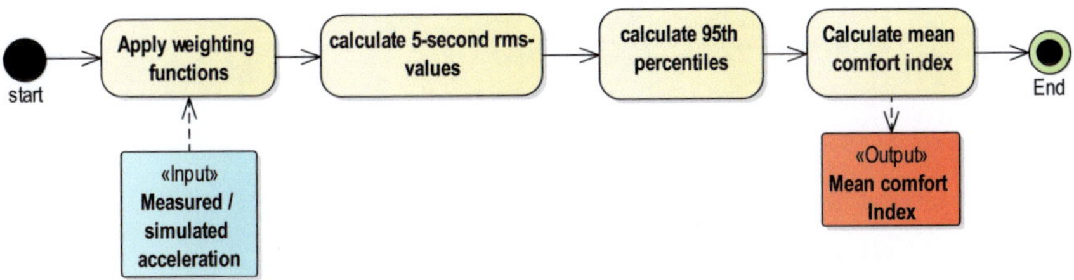

Figure 76:   **Mean comfort index N_{MV} calculation process (own work)**

Accelerations (a_{xP}, a_{yP} and a_{zP}) are obtained through accelerometers placed on the floor of the vehicle for simulations at a constant speed for 5 minutes. The sub-indices x, y and z mean the direction in the coordinate system and the P means the position of the accelerometer which in this case is the floor.

In this example that follows, accelerations correspond to a simulation performed at 50 km/h on track 330i at its worst track geometry (i.e. day 320).

The measured accelerations are first filtered with weighting curves that account for different degrees of sensitivity displayed by different individuals as a function of frequency (The British Standards Institution 2009). Weighting curve W_d is used for accelerations a_{xP} and a_{yP} resulting in weighted accelerations $a_{xP}^{W_d}$ and $a_{yP}^{W_d}$ respectively. Weighting curve W_b is used for acceleration a_{zP} resulting in weighted acceleration $a_{zP}^{W_b}$.

As mentioned in (The British Standards Institution 2009), the overall frequency weighting function W_d is a product of a high and a low filter $H_h(f) \cdot H_l(f)$ (i.e. band-limiting filter) and an acceleration velocity transition filter $H_t(f)$, while the overall weighting function W_b is the product of the filters already mentioned plus an upward step filter $H_s(f)$ which is proportional to the jerk. For more information about the transfer function refer to section C.2 of standard EN 12299 (The British Standards Institution 2009). Below are the relationships that form the overall transfer functions of weighting curves W_d and W_b.

Overall transfer function of weighting curve W_d

$$H_{W_d}(f) = H_h(f) \cdot H_l(f) \cdot H_t(f) \tag{39}$$

Overall transfer of weighting curve W_b

$$H_{W_b}(f) = H_h(f) \cdot H_l(f) \cdot H_t(f) \cdot H_s(f) \tag{40}$$

where

$H_{W_d}(f)$ = Overall weighting function W_d

$H_{W_b}(f)$ = Overall weighting function W_b

$H_h(f)$ = High pass filter

$H_l(f)$ = Low pass filter

$H_t(f)$ = Acceleration to velocity transition filter

$H_s(f)$ = Upward step filter (proportional to the jerk)

Figure 77 depicts the weighting curves used to filter accelerations.

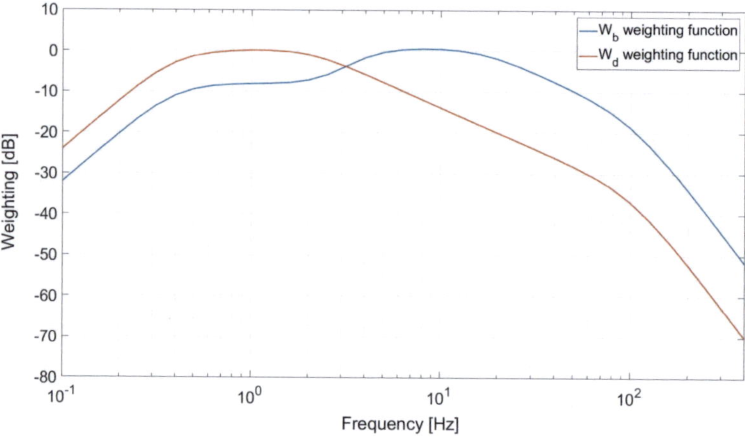

Figure 77:   Weighting curves W_d and W_b (own work)

Figure 78 to Figure 80 show simulated and weighted accelerations.

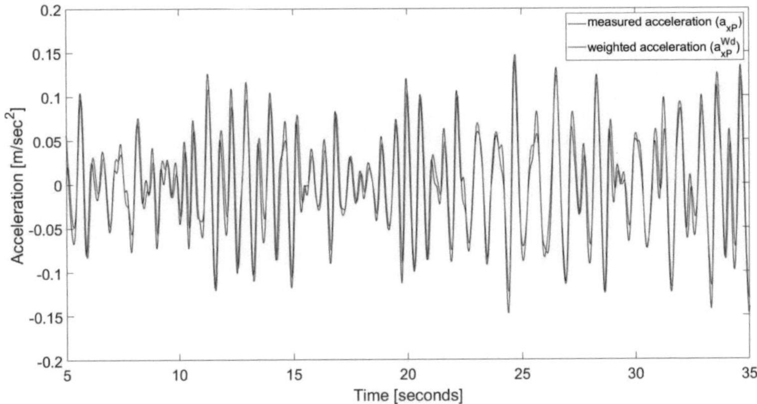

Figure 78: Simulated a_{xP} and weighted $a_{xP}^{W_d}$ accelerations in the x-direction (own work)

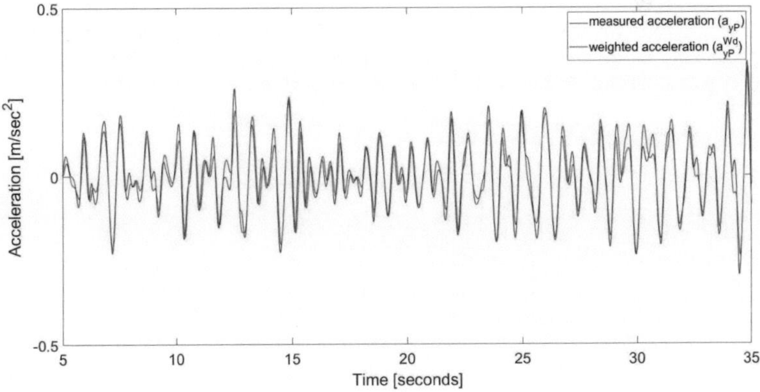

Figure 79: Simulated a_{yP} and weighted $a_{yP}^{W_d}$ accelerations in the y-direction (own work)

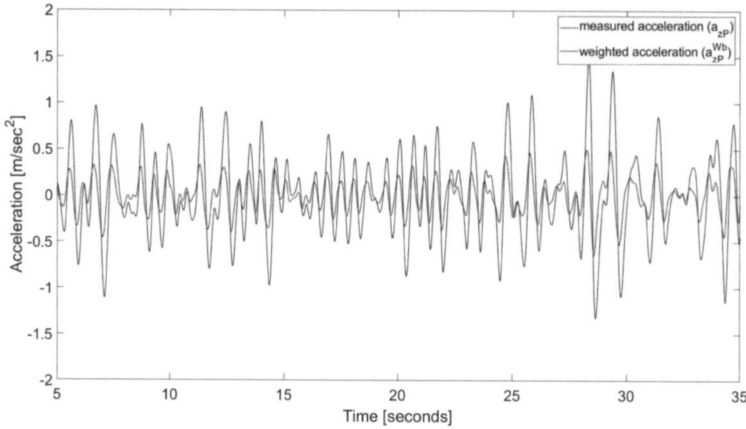

Figure 80: Simulated a_{zP} and weighted $a_{zP}^{W_b}$ accelerations in the z-direction (own work)

To complete the evaluation process, the weighted signals ($a_{xP}^{W_d}$, $a_{yP}^{W_d}$, $a_{zP}^{W_b}$) undergo a RMS evaluation of five-second intervals and a 95th percentile calculation ($a_{xP95}^{W_d}$, $a_{yP95}^{W_d}$, $a_{zP95}^{W_b}$). The results are shown in Table 29 and Figure 81 to Figure 83.

Acceleration (weighted, RMS, 95th percentile)	[m/s²]
$a_{xP95}^{W_d}$	0.0796
$a_{yP95}^{W_d}$	0.1270
$a_{zP95}^{W_b}$	0.2633

Table 29: Results of acceleration weighting, RMS evaluation and 95th percentile calculation

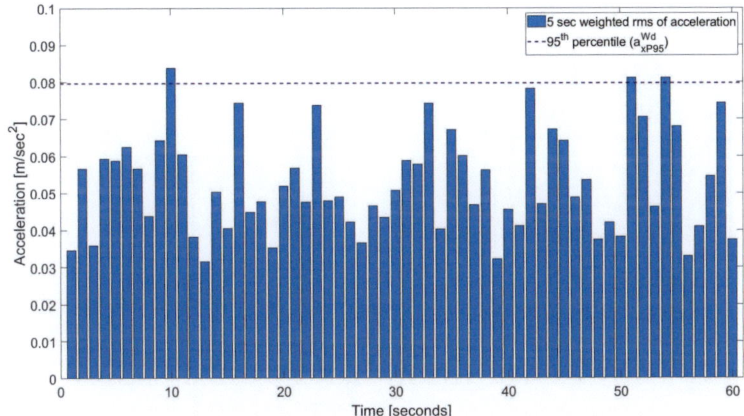

Figure 81:   RMS and 95th percentile of weighted acceleration $a_{xP95}^{W_d}$ in the x-direction (own work)

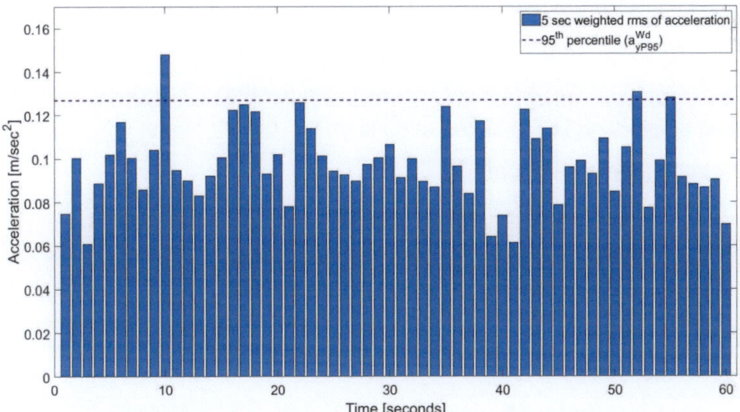

Figure 82:   RMS and 95th percentile of weighted acceleration $a_{yP95}^{W_d}$ in the y-direction (own work)

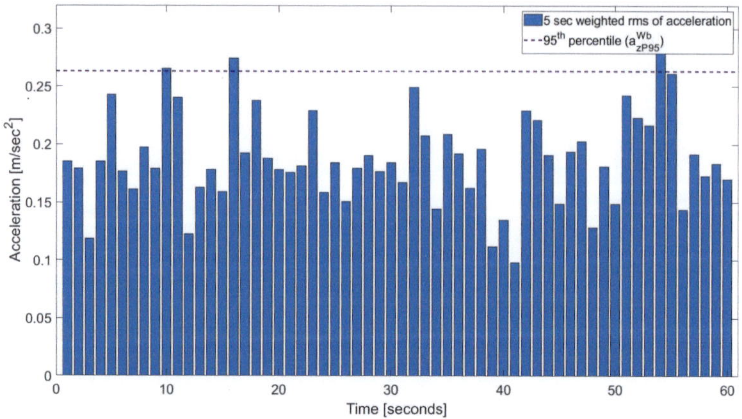

Figure 83: **RMS and 95th percentile of weighted acceleration $a^{W_b}_{zP95}$ in the z-direction (own work)**

The values obtained are used in the following formula (same as Equation 5) to calculate the mean comfort index N_{MV} which results in value of 1.82. EN 12299 provides a perception scale that can be used to evaluate the mean comfort value (see Table 30). The mean comfort index calculated at 50 km/h for day 320 corresponds to a comfortable ride zone (see also Figure 63).

$$N_{MV} = 6\sqrt{(a^{W_d}_{XP95})^2 + (a^{W_d}_{YP95})^2 + (a^{W_b}_{ZP95})^2} \tag{41}$$

$N_{MV} < 1.5$	Very comfortable
$1.5 \leq N_{MV} < 2.5$	Comfortable
$2.5 \leq N_{MV} < 3.5$	Medium
$3.5 \leq N_{MV} < 4.5$	Uncomfortable
$N_{MV} \geq 4.5$	Very comfortable

Table 30: **Scale for the mean comfort index N_{MV} per (The British Standards Institution 2009)**

Appendix IV: Track Data Provided and Nomenclature

The data of 11 inspection dates were provided for several track sections to perform this study. However, tracks 300 and 400 were used in its development. Table 31 and Table 32 present the details of the tracks and nomenclature used throughout this study.

Track 300				
Track 330i (inbound) = (300i)			Track 330o (outbound) = (300o)	
Inspection Date	Day in study	Quality in TGI plot	Observations	History
05.22.2013	0	$Q_{1(300)}$		Before renewal
04.07.2014	320	$Q_{2(300)}$	Worst TG observed	
09.29.2014	495	$Q_{3(300)}$		After renewal
03.26.2015	673	$Q_{4(300)}$	Best TG observed	
10.13.2015	872	$Q_{5(300)}$		
03.04.2016	1037	$Q_{6(300)}$		
09.15.2016	1212	$Q_{7(300)}$		
04.03.2017	1412	$Q_{8(300)}$		
09.20.2017	1582	$Q_{9(300)}$		
13.03.2018	1756	$Q_{10(300)}$		

Table 31: **Track 300 details**

Track 400			
Track 400i (inbound) = (400i)		Track 400o (outbound) = (400o)	
Inspection	Day in study	Quality in TGI plot	History
05.22.2013	0	$Q_{1(400o)}$	Not available / not provided
17.09.2013	118	$Q_{2(400o)}$	
04.07.2014	320	$Q_{3(400o)}$	
29.09.2014	495	$Q_{4(400o)}$	
26.03.2015	673	$Q_{5(400o)}$	
13.10.2015	874	$Q_{6(400o)}$	
24.03.2016	1037	$Q_{7(400o)}$	
21.09.2016	1218	$Q_{8(400o)}$	
12.04.2017	1421	Data not available	
19.09.2017	1581	$Q_{10(400)}$	
06.03.2018	1751	$Q_{11(400)}$	

Table 32: **Track 400 details**

Appendix V: Synthetic Signal Generation

As discussed in section 3.4.1, synthetic signals require a minimum number of samples and frequency steps to produce random signals that represent a signal per its PSD and statistical characteristics. The conclusion is that random signals should be generated with a minimum of $N_s = 100$ samples and $N = 1500$ frequency steps. Figure 84 to Figure 86 show the representative statistics of synthetic signals generated with different number of samples per signal.

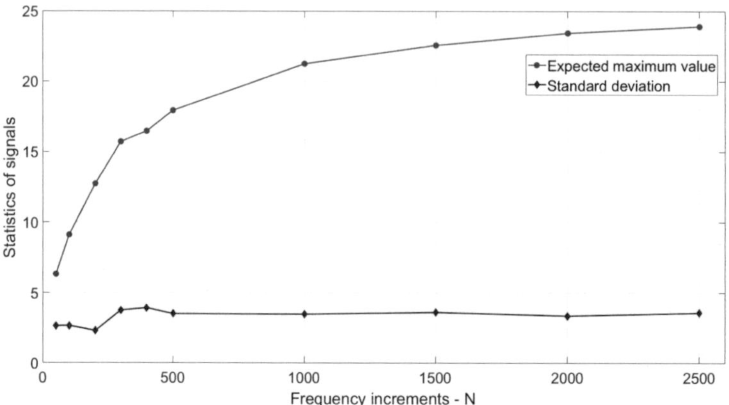

Figure 84:   Statistics of signals for N_s = 50 (own work)

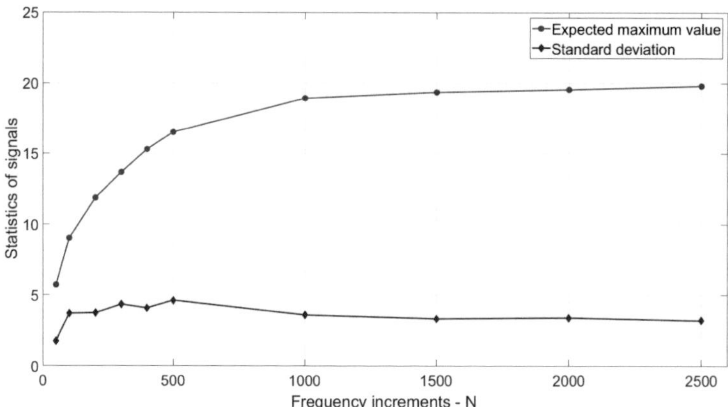

Figure 85:   Statistics of signals for N_s = 100 (own work)

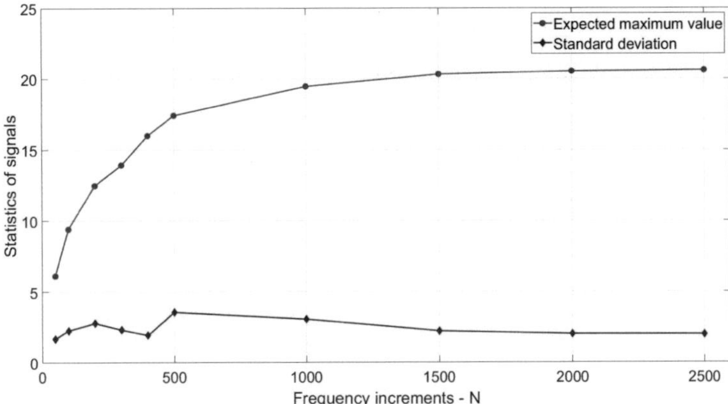

Figure 86:   **Statistics of signals for N_s = 300 (own work)**

As mentioned in section 3.4.1, a signal with an $N_s = 50$ does not reach a stable behavior for the expected maximum value at the maximum 2500 frequency steps.

Appendix VI: Stiffness and Damping Calculations

Appendix VI.A: Calculation of track stiffness

The track stiffness calculation in this section is based on track total modulus u per (Selig and Waters 1994). According to (Berggren April /2009) u is defined as the vertical foundation supporting force per unit length (q) per unit deflection (δ):

$$u = \frac{q}{\delta} \tag{42}$$

where

u = track modulus   [MPa]

q = force per unit length   [N/m]

δ = deflection   [mm]

To calculate the stiffness (k) of the track form the track modulus the following formula is used:

$$u = \frac{k^{1/3}}{(E_r I_r)^{1/3}} \tag{43}$$

where

k = stiffness of the track   [N/m]

E_r = modulus of elasticity of the rail   [N/cm^2]

I_r = moment of inertia of the rail   [cm^4]

Rearranging the equation:

$$k = \sqrt[4]{(64\,E_r\,I_r)\,u^3} \qquad\qquad \text{[N/m]} \tag{44}$$

In (Selig and Waters 1994), it is mentioned that a track modulus u = 28 MPa corresponds to a minimum value for a continuous good track condition, which according to (Bezin et al. 2009) corresponds to a stiffness of k = 55 kN/mm for a conventional line with a rail type UIC60. I n (Li et al. 2016) a range of track modulus is provided for values in MPa for soft medium and stiff soil conditions which correspond respectively to poor, good and very good track conditions. The range of stiffness values for an LRT rail 49E1 with an E_r = 21 x 10^6 N/cm^2 and an I_r = 1816 cm^4 per standard DINS49 are shown in Table 33.

Soil condition / Structural Track quality	Track modulus [MPa]	Track stiffness [kN/mm]
Soft / poor	7 - 8	17 - 48
Medium / good	28 - 69	48 - 94
Stiff / very good	69 - 138	94 - 159

Table 33: **Calculation of track stiffness for LRT rail 49E1 based on track modulus u**

The results agree with the range of values the stiffness of a good quality of track provided in (Lichtberger 2005a) of 50 – 100 kN/mm.

Appendix VI.B: Calculation of track damping

Track damping calculation based on total track modulus C per (Göbel and Lieberenz 2004; Rapp 2017). In this study the value of C was taken as 100 MN/m^3 in accordance to the discussion regarding the structural quality of the track assumed (see section 4.1). In (Rapp 2017) the relation between the total track modulus C and the track modulus of the ground C_g is established through the formula below, which has previously discounted the track modulus of sleepers, sole of the sleeper and ballast mat as shown in (Rapp 2017):

$$C_g = \frac{1}{{}^1/_C - {}^1/_{C_{rv}}} \qquad \text{[MN/m}^3\text{]} \quad (45)$$

were

C_{rp} = the track modulus of the rail pad calculated with the following relation:

$$C_{rp} = {}^{K_{rp}}/_{(A_s/2)} \qquad \text{[MN/m}^3\text{]} \quad (46)$$

where

K_{rp} = the rail pad stiffness in [MN/m]

A_s = effective area of sleeper (the area of pressure exerted by the sleeper minus the pressure free area in the middle)

In (Giannakos 2010) K_{rp} is reported to vary between 48 – 600 MN/m . The value of 386 MN/m used in the calculation of the C_{rp} in this work was provided in (Rapp 2017). A value of A_s = 0.57, used in calculations was obtained from (Rapp 2017). The value of the track modulus of the ground calculated is C_g = 108 MN/m^3. This value is used to calculate the wet density (ρ_w), the Poisson ration (v) and the shear modulus (G) of the underlaying soil required to find the damping of the ground (D_g) as shown by the formula below

$$D_g = \frac{3.4 \cdot r_0 \cdot 2}{(1-v)} \cdot \sqrt{\rho \cdot G \cdot 10^9} \qquad \text{[Ns/m]} \qquad (47)$$

where

r_0 = idealized radius of pressure influence cone calculated as follows

$$r_0 = \sqrt{(2\,L\,b_l/\pi)} \qquad \text{[m]} \qquad (48)$$

where

$$L = \sqrt[4]{\frac{4\,E_r I_r}{C\,b_l}}$$

| Elastic modulus of the rail E_r = 2.1 x 10^7 [N/cm²] |
| Moment of inertia of the rail I_r = 1816 [cm⁴] |

were

b_l = width of the hypothetical long sleeper given in calculated as follows

$$b_l = A_s/(2 \cdot a) \qquad \text{[m]} \qquad (49)$$

where

a = distance between two sleepers on center given in [m] (0.65 m)

The values of the wet density (ρ_w) and Poisson ratio (v) depending on the track modulus of the ground (C_g), are calculated through the empirical formulas developed in (Rapp 2017):

$$\rho_w = 2.05 \cdot C_g^{0.1} - 1.191 \qquad C_g \;\epsilon\; (20,\,245) \qquad \text{[g/cm}^3\text{]} \quad (50)$$

$$v = 1.362 \cdot C_g^{-0.142} - 0.4 \qquad C_g \;\epsilon\; (20,\,245) \qquad \text{[-]} \quad (51)$$

The shear modulus (G) is calculated with the following formula:

$$G = C_g \cdot (1-v) \cdot \sqrt{\pi/8} \cdot \sqrt{L \cdot b_l} \qquad \text{[MN/n}^2\text{]} \quad (52)$$

Results for different structural qualities, i.e. 50 MN/m^3 (poor quality), 100 MN/m^3 (good quality) and 150 MN/m^3 (very good quality) are presented in Table 21 in section 5.1.

Appendix VII: Vehicle Data

The following values were provided by SSB to IMA to build the model as close as possible to reality. The data presented follows values obtained from (Holzscheiter 2017; Bauer and Theurer 2000).

Description	Value
Length	38,150 [mm]
Width	2,650 [mm]
Weight (empty)	55.6 [tons]
Weight (allowed)	23 [tons]

Table 34:   **Properties of vehicle DT8.10 (Bauer and Theurer 2000)**

Description	Mass [kg]
Cabin mass	16,672
Bolster	350
Bogie	3,200
Wheelset	781

Table 35:   **Mass of individual components of the DT8.10 vehicle per (Holzscheiter 2017)**

Description	Direction	Stiffness [N/m]	Damping [Ns/m]
Primary spring	x	60×10^6	6,000
	y	4.622×10^6	4,000
	z	1.760×10^6	0
Primary damper	z	0	36,000
Secondary spring	x	108,000	0
	y	108,000	0
	z	343,000	0
Secondary damper	y	0	21,000
	z	0	30,000
Traction rod		5×10^8	400,000

Table 36:   **Values of suspension components of the DT8.10 per (Holzscheiter 2017)**

Range	Stiffness [N/m]	Damping [Ns/m]
0 – 5 mm	0	0
5 – 25 mm	650,000	0
> 25 mm	1×10^8	0

Table 37:   **Stifness of the Queranschlag Rollenkranzträger**

Appendix VIII: Location of Sensors

According to standard EN 12299 (The British Standards Institution 2009) accelerations at a given point in a vehicle are closely dependent upon the location of that point. Hence measurements shall be carried out at the center of the body and at both ends of the passenger compartment, at the seats most closely located to these positions. Furthermore, standard EN 12299 (The British Standards Institution 2009) mentions that accelerometers shall be fixed to the floor as closely as possible to the vertical projection at the center of the seat pan (preferably less than 100 mm from this point). Figure 87 shows the positions of accelerometers for the calculation of the mean comfort index in this study.

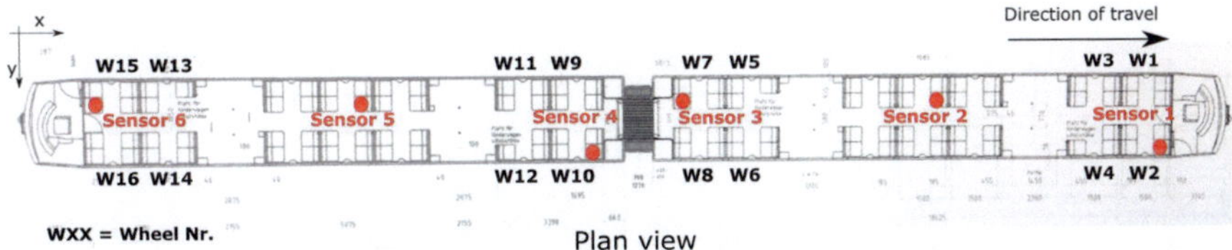

Figure 87: **Position of sensors in the cabin per (The British Standards Institution 2009) (own work based on image form (Bauer and Theurer 2000; Holzscheiter 2017))**

Simulation results showed that the highest accelerations were experienced by sensors 6 (S6) and 1 (S1) away from the center of mass of the cabins and train consist. To study the effect that the suspension has on the measured acceleration inside the cabin, axle box accelerometers were placed at the wheels close to S6 (highest accelerations along with S1), i.e. wheels 15 (W15) and 16 (W16) as seen in Figure 88.

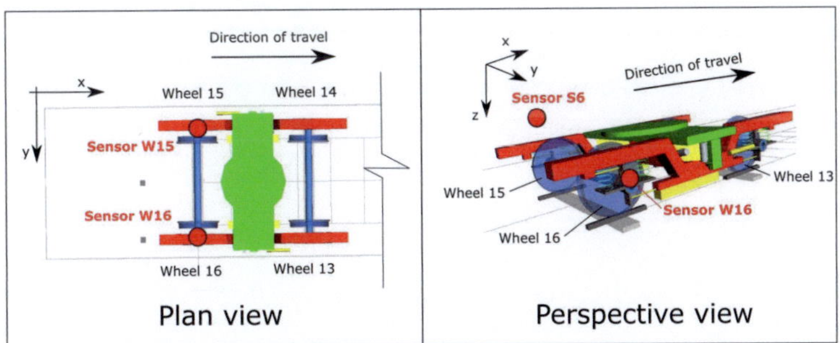

Figure 88: **Location of sensors at the axle box of wheels 15 and 16 (own work based on screenshot of MBS model)**

Appendix IX: Real vs Simulated Vertical Accelerations

To compare real vs. simulated accelerations, eight runs on track 330 inbound and outbound were conducted. Real accelerations were measured inside the passenger cabin at the location of sensor 6 (see Appendix VIII: Location of Sensors). The measured accelerations (Figure 89) are treated to eliminate excessive noise.

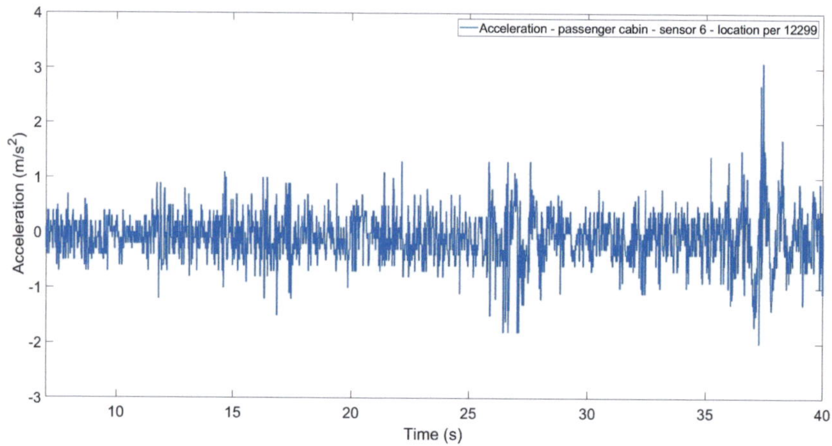

Figure 89:   **Unfiltered accelerations measured in real train at speeds averaging lower than km/h (own work)**

Figure 90 depicts the process followed to determine measured accelerations.

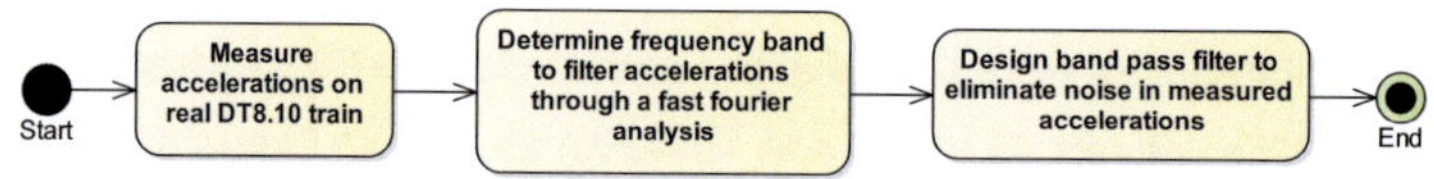

Figure 90:   **Process to treat accelerations measured on real light rail vehicle in Stuttgart (own work)**

Figure 91 shows the treated acceleration which are later compared to simulated SIM-PACK© accelerations as seen in Figure 92 to Figure 99.

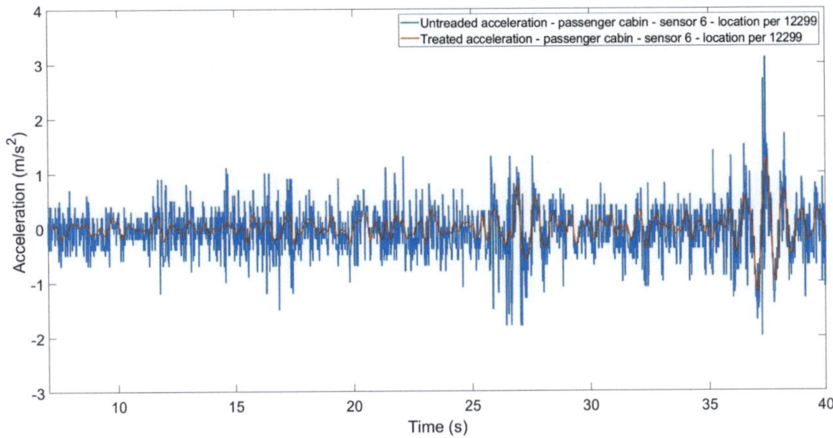

Figure 91: **Comparison between filtered and unfiltered acceleration measured on moving train for a speed average lower than 50 km/h (own work)**

As it can be seen in the graphs below (Figure 92 to Figure 99), the accelerations measured are within the same order of magnitude as the simulated accelerations in SIM-PACK©. However, as can also be observed, the magnitudes of the measured accelerations are in general lower. As explained in section 5.1, the differences between the measured and simulated accelerations might lay in the differences in speeds (simulated speed is 50 km/h while the speed of the measured train was in general below 50 km/h with averages ranging between 34 km/h and 48 km/h), the stiffness of the track, the presence of discrete events on the real track not present on the simulated track, differences in temperature, condition of the vehicles used for the measurements, etc.

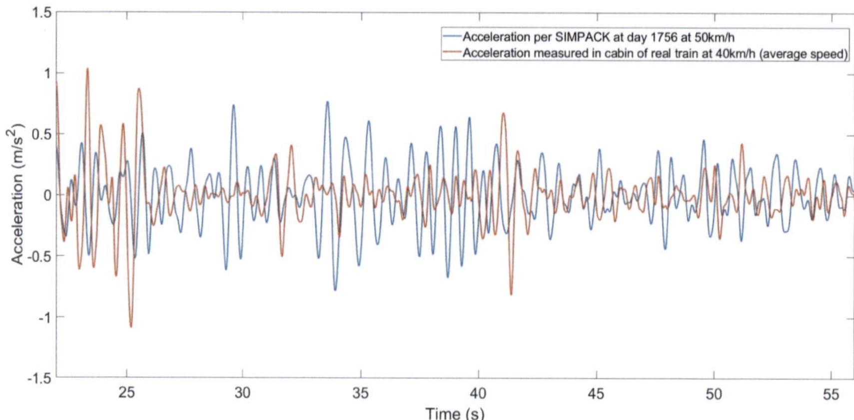

Figure 92:   **Measured acceleration in train at 40 km/h (average speed) vs acceleration in SIMPACK©
at 50 km/h (own work)**

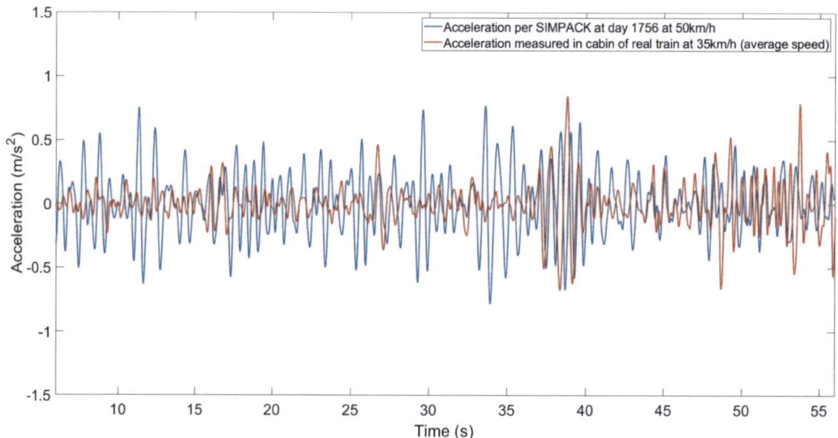

Figure 93:   **Measured acceleration in train at 35 km/h (average speed) vs acceleration in SIMPACK©
at 50 km/h (own work)**

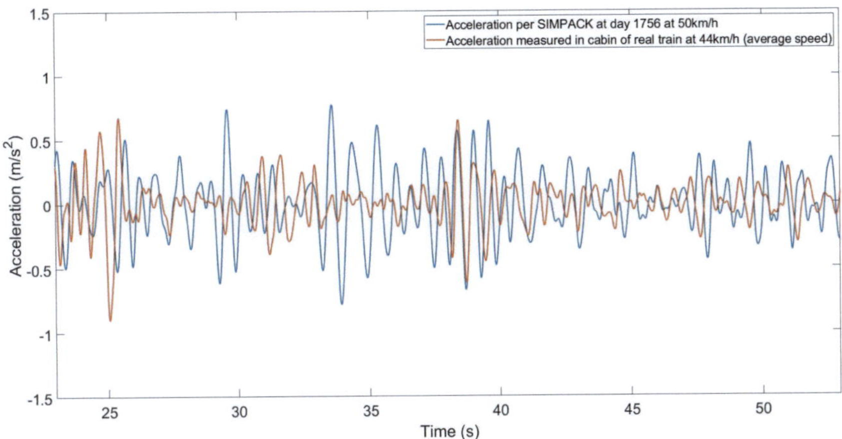

Figure 94:   **Measured acceleration in train at 44 km/h (average speed) vs acceleration in SIMPACK©
at 50 km/h (own work)**

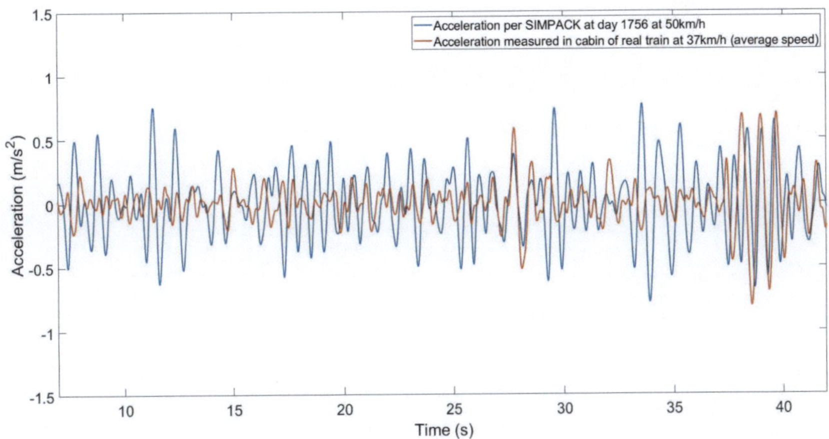

Figure 95:   **Measured acceleration in train at 37 km/h (average speed) vs acceleration in SIMPACK©
at 50 km/h (own work)**

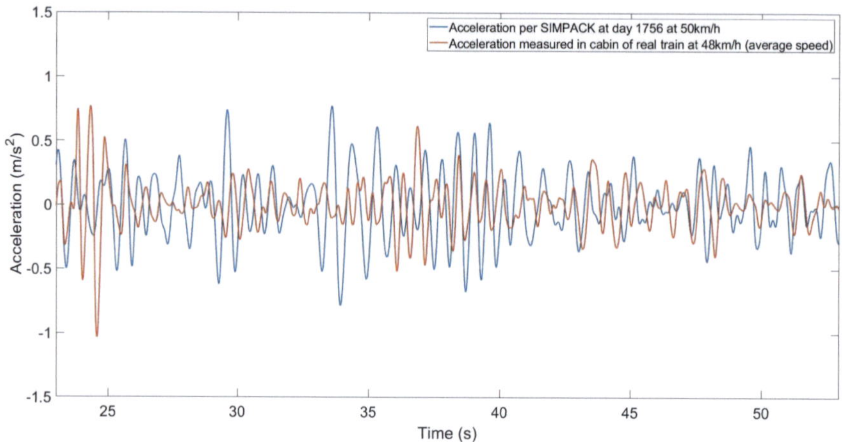

Figure 96: **Measured acceleration in train at 48 km/h (average speed) vs acceleration in SIMPACK©**
at 50 km/h (own work)

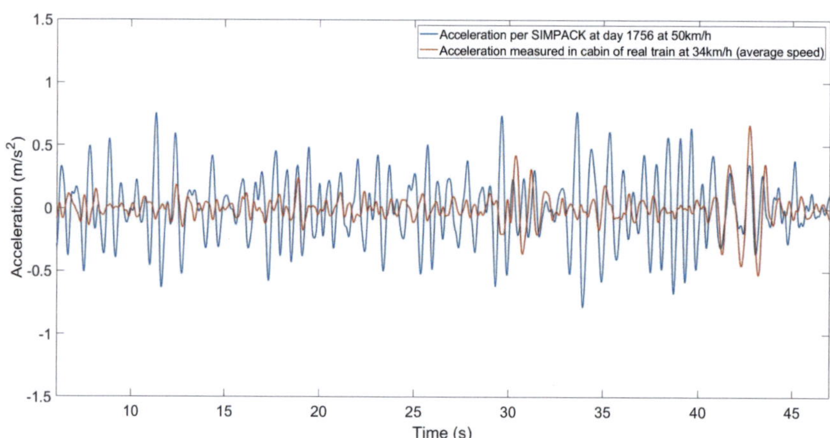

Figure 97: **Measured acceleration in train at 34 km/h (average speed) vs acceleration in SIMPACK©**
at 50 km/h (own work)

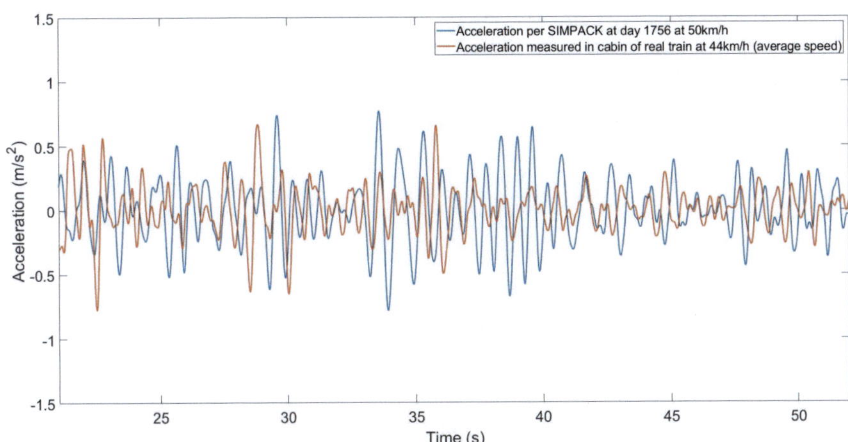

Figure 98: **Measured acceleration in train at 44 km/h (average speed) vs acceleration in SIMPACK© at 50 km/h (own work)**

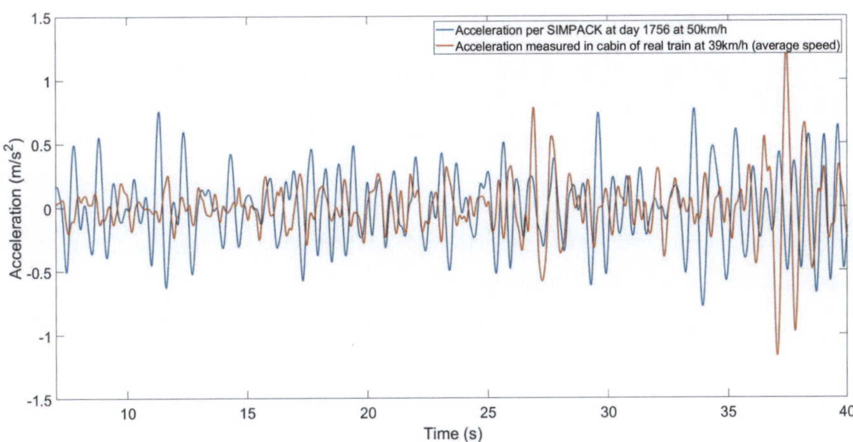

Figure 99: **Measured acceleration in train at 39 km/h (average speed) vs acceleration in SIMPACK© at 50 km/h (own work)**

Appendix X: Vertical Stress Calculation

Appendix X.A: Calculation of stress for vehicle response based on measured track data and calculation of allowable stress

The vertical load is treated per EN 14363 (The British Standards Institution 2016). The maximum dynamic load observed (per speed simulated) is used to calculate the maximum stress under the sleeper at the given speed, which in turn is used to determine stresses exerted on the track at different depths (e.g. subgrade) according to the elastic two layer system theory by Burmeister as shown in (Göbel and Lieberenz 2004). The vertical stress depends on the strength (modular) ratio E_1/E_2 where E_1 is the modulus of deformation of the upper layer (ballast and sub-ballast) while E_2 is the modulus of deformation of the lower layer (i.e. subgrade) (Punmia and Jain 2005).

Below is the process to calculate the allowable stress σ_z on the subgrade directly under one sleeper for a track using a rail type 49E1, a sleeper type B70 and ballast and sub-ballast layer of minimum 0.25m each. The calculation is performed for a "hard" layer of 50 cm and 60 cm to account for tracks that might provide the minimum thickness required by SSB. The stress calculated is compared to the allowable stress at the subgrade.

To calculate the stress, it is necessary to calculate first the maximum load that the track might experience using the following relationship:

$$Q_{max} = Q \cdot (1 + P \cdot SD) \tag{53}$$

where

Q = observed load   [KN]

SD = standard deviation   [KN]

P = probability factor[18]   [-]

---

[18] Probability factor to determine the maximum statistically related safety value per [Göbel and Lieberenz (2004)]

SD is calculated with the following relations depending of the type of railway vehicle and the speed range of the train in accordance to table 2.11 in (Göbel and Lieberenz 2004).

$$SD = n \cdot \varphi \tag{54}$$

where

n = factor depending on track and track geometry quality $\qquad$ [-]

φ = velocity coefficient per Table 38 $\qquad$ [-]

Speed V_{veh} [km/h]	Velocity coefficient [-]
$V_{veh} \leq 60$	1
Passenger $60 < V_{veh} \leq 300$	$1 + 0.5 \cdot \dfrac{(v - 60)}{190}$

Table 38:   Velocity coefficients per (Göbel and Lieberenz 2004)

Then the Q_{max} is used to calculate the maximum stress p_{max} under the sleeper as follows:

$$p_{max} = \frac{Q_{max}}{2 \cdot b_l \cdot L} \tag{55}$$

where

b_l = width of the hypothetical long sleeper (Appendix VI: Stiffness and Damping Cal-

L = characteristic length (Appendix VI: Stiffness and Damping Calculations)

Then the Burmeister's influence coefficient k_s needs to be calculated based on the following relationship:

$$k_s = 1 - \left(1 - \bar{k}_s\right) \cdot (z/h)^2 \tag{56}$$

where

$\bar{k}_s$ = correction value per Figure 100:   Correction value $\bar{k}_s$ for the Burmeister influ-

z = depth $\qquad$ [m]

h = depth of "hard" layer $\qquad$ [m]

Finally, the vertical stress is calculated with the following formula:

$$\sigma_z = p_{max} \cdot i_z \cdot k_s \tag{57}$$

where

p_{max} = Maximum vertical stress   [N/cm^2]

i_z = stress influence value per Figure 102   [-]

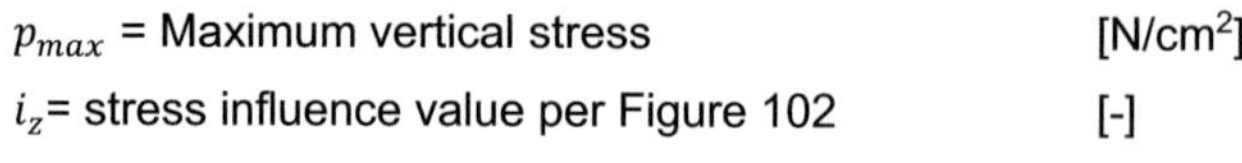

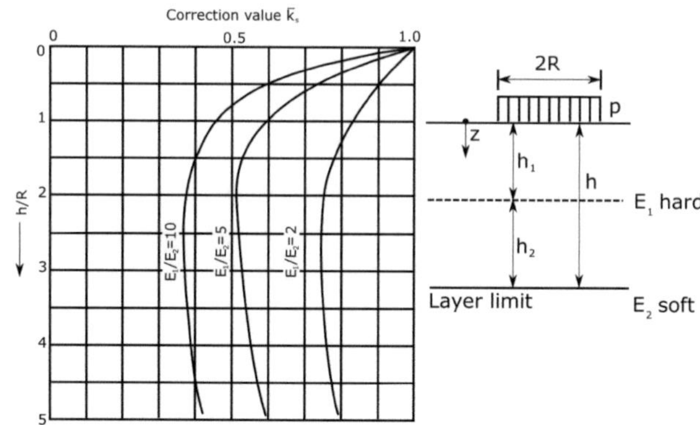

Figure 100: **Correction value $\bar{k}_s$ for the Burmeister influence coefficient k_s per (Göbel and Lieberenz 2004)**

The radius of the loaded circular area to determine the correction value in Figure 100 depends on the sleeper's effective area and is calculated with the following formula:

$$R = \sqrt{A_s/(2 \cdot \pi)} \tag{58}$$

where

A_s = effective area of sleeper   [cm^2]

The depth h is the sum of the depth of ballast layer h_1 and sub-ballast layer h_2.

To calculate the effect of the maximum load Q_{max} at different positions away from the concentrated load, the relationships dependent on the radius R are used shown in Figure 101. Each relationship corresponds to a curve number in the stress influence value shown in Figure 102. In the case under study, the stress is calculated directly under Q_{max} which corresponds to curve number 1 in Figure 102.

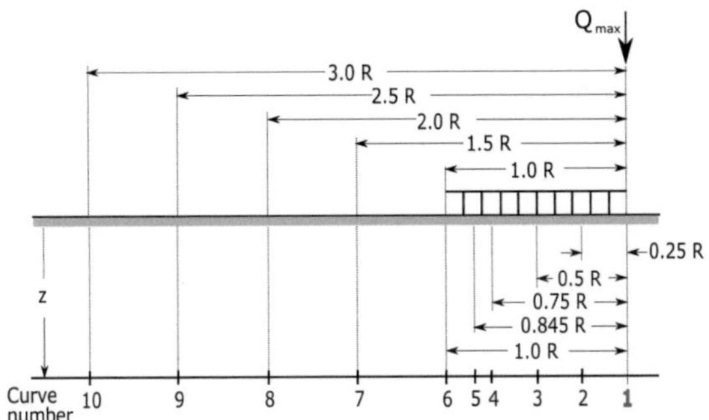

Figure 101:   **Load influence from the point of maximum load Q_{max} per TU Darmstadt, Institut für Geotechnik**

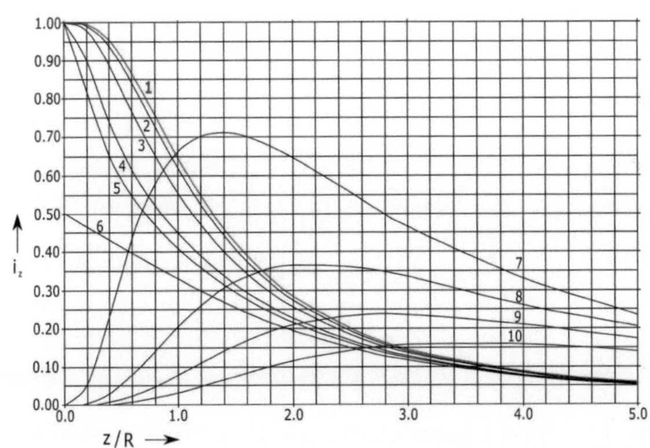

Figure 102:   **Stress influence value i_z for a circular type loading per TU Darmstadt, Institut für Geotechnik**

To compare the stress calculated, the maximum allowable stress on the subgrade of a main line is calculated following the approach in (Schmutz 1980) with the following formula:

$$\sigma_{allow} = 1.64 \ x \ 10^{-3} \cdot E_{sub} \tag{59}$$

where

E_{sub} = Modulus of the subgrade   (also E_2)   [N/cm²]

The following values are taken to perform the calculations of the maximum stress at 50 and 60 cm depth for a vehicle traveling at 80 km/h (calculation performed in MATLAB©:

E_1 = 15,000 [N/cm²]

E_{sub} = E_2 = 1,500 [N/cm²]

A_s = 5700 [cm²]

V_{veh} = 80 [km/h]

n = 0.15 (good)

P = 1.65

b_l = 43.84 [cm] per Appendix VI: Stiffness and Damping Calculations

L = 76.80 [cm] per Appendix VI: Stiffness and Damping Calculations

The stress at 50 and 60 cm depths are compared to the allowable stress as shown in Table 39. As it can be seen the pressures exerted by the LRV for the provided track geometry data for track 330i at its worse quality (day 320) for the highest speed possible (80 km/h) do not surpass or get close to the allowable subgrade pressure.

Depth [cm]	Stress calculated σ_z [N/cm²]	Allowable stress σ_{allow} [N/cm²]
50	1.50	2.46
60	1.15	

Table 39:   Subgrade stress σ_z at 50 and 60 cm and allowable stress σ_{allow} at subgrade (own work)

Similarly, stresses at different speeds were calculated. The results are shown in Table 23 in section 5.2.

However, as the magnitudes of the PSD irregularities for the worse days observed for tracks 300 and 400 are increasing to worsen the track geometry (see section 5.3), it is necessary to check the stress level at these new higher irregularities. The following appendix calculates the new stresses and compares them to the allowable stress to see if they remain acceptable or determine a limit of intervention based on track structural deterioration.

Appendix X.B: Calculation of stress for vehicle responses based on data with increased magnitudes and comparison to allowable stresses at different depths

For allowable stress for increased signal magnitudes see Table 45 in Appendix XII: Magnitude Increase "Fine-Tuning".

Appendix XI: Track Parameter Magnitude Increase

This appendix is a summary of the approach to determine an appropriate factor increase for the parameters of measured signals to create worse track geometrical qualities following the study presented in (Araji 2018).

The first step consisted in assuming that the vertical and horizontal irregularities dominate track geometrical quality and hence comfort level. The process to determine the appropriate increase factor was iterative and in the frequency domain (i.e. applied to PSD of the signals). In the initial state all parameters (i.e. vertical, lateral, cross level and gauge) were taken as a factor of one, which corresponds to the PSD input function for the worst track quality observed. From the initial state, the different parameters were scaled gradually and their influence on the comfort level noted.

To determine the influence of the horizontal irregularity, the horizontal (H) parameter was increased gradually in magnitude while keeping the other parameters constant as seen in Table 40.

Factor	Average comfort	% change from previous value	% change from initial state
H1	1.74	0.00	0.00
H2	1.95	12.29	12.29
H3	2.04	4.43	17.27
H4	2.22	8.95	27.76
H5	2.31	4.07	32.96
H6	2.44	5.61	40.43

Table 40: **Effect of lateral irregularities on comfort level per (Araji 2018)**

The result was that the horizontal increase in magnitude produced a change in comfort of a maximum of 40%. A similar exercise was performed on the vertical (V) irregularity, finding that the change in comfort level was up to 103% as it can be seen in Table 41.

Factor	Average comfort	% change from previous value	% change from initial state
V1	1.74	0.00	0.00
V2	2.23	28.71	28.71
V3	2.69	20.16	54.66
V4	3.00	11.84	72.97
V5	3.38	12.47	94.55
V6	3.53	4.48	103.27

Table 41: **Effect of vertical irregularity on comfort level per (Araji 2018)**

As it can be observed, an uncomfortable level (i.e. comfort index 3.5) is reached by only scaling the vertical irregularity. This occurs at a factor increase of 6; hence, magnitude 6 was kept constant while lateral irregularities were gradually increased in order to see their influence on the vehicle reaction. From Table 42, it is evident that an increase of lateral irregularities causes little impact on the value of the comfort level. This suggests that the vertical irregularity influences the most the vehicle reaction for the track/vehicle system under study (Araji 2018).

Factor	Average comfort	% change from previous value	% change from initial state
V6H1	3.53	0.00	103.27
V6H3	3.71	5.22	113.88
V6H4	3.70	-0.46	112.89
V6H5	3.74	1.32	115.71
V6H6	3.80	1.58	119.12

Table 42: **Horizontal magnitude increase while maintaining the vertical irregularity constant at 6 per (Araji 2018)**

Likewise, it was determined that cross level and gauge do not exert a large influence regarding comfort deterioration either.

The next step consisted in increasing vertical irregularities to a magnitude of 10 while keeping the gauge and cross level at magnitude 1 and, arbitrarily, the lateral irregularity constant at magnitude 3.

Factor	Average comfort	% change from previous value	% change from initial state
H3V1	2.04	0.00	17.27
H3V2	2.53	24.22	45.67
H3V3	2.82	11.32	62.16
H3V4	3.21	14.04	84.93
H3V5	3.44	7.30	98.43
H3V6	3.71	7.79	113.88
H3V7	3.86	3.94	122.31
H3V8	4.07	5.50	134.53
H3V9	4.25	4.48	145.03
H3V10	4.49	5.59	158.72

Table 43: **Increase in vertical irregularity while keeping horizontal irregularities at 3 per (Araji 2018)**

This exercise resulted in magnitudes that got close to the limits of comfort sought. Table 43 shows that the limit of comfort / intervention was almost reached at H3V5, while a limit of very uncomfortable was reached approximately at H3V10. Since it was known that the comfort level is reached at H5, gauge and cross level irregularities were added, first separately and then combined at different magnitudes. Several trials were performed to assess whether it still made sense to increase the magnitude of the horizontal irregularity.

In conclusion, the effect of having varying magnitudes for gauge, cross level, and lateral irregularity seemed to have minimal effect on the value of the comfort (±1% to 2% change) which is solely dominated by the effect of the vertical irregularity. Therefore, it made sense to equalize the order of magnitude for all irregularities. As it can be seen in Table 44, the comfort level was approached at H5V5G5CL5, while the very uncomfortable level was approached at H9V9G9CL9. An alert limit was approached at H3V3G3CL3. However, the alert limit value was taken from the magnitude 3 of the signals measured on track 330 at day 320 which represented the worse track geometry quality measured on that track.

Factor	Average comfort	% change from previous value	% change from initial state
H2V2G2CL2	2.37	0.00	36.30
H3V3G3CL3	2.81	18.88	62.04
H3V5	3.44	22.46	98.43
H4V5	3.53	2.41	103.21
H5V5	3.52	-0.09	103.02
H5V5G5CL5	3.51	-0.47	102.06
H3V5G2	3.41	-2.79	96.41
H3V5CL2	3.44	0.97	98.32
H3V5G2CL2	3.41	-1.00	96.33
H3V5G3CL3	3.46	1.42	99.13
H5V10	4.45	28.87	156.61
H5V10G5CL5	4.46	0.07	156.79
H9V9G9CL9	4.49	0.65	158.45

Table 44: **Increase of magnitudes for all parameters for different values and equal values for all parameters per (Araji 2018)**

Finally, (Araji 2018) reported that the rate of change in vertical irregularities is larger than for lateral irregularities. This is in accordance with the spatial – frequency studies performed by (Álvarez 2018), which characterized the track under study reporting that vertical irregularities present the higher degree of defects and higher deterioration rate.

Appendix XII: Magnitude Increase "Fine-Tuning"

Table 45 shows the increase factors that parameters for day 0 of track 400 were subjected to in order to obtain the required comfort and safety limits.

Track / Day	Factor	Comfort Index [-]	Comfort index limits	Y/Q [-]	Y/Q limit [-]	Q [kN]	Q_{max} [kN]	σ_z [N/cm²] 50 [cm]	σ_z [N/cm²] 60 [cm]	σ_{allow} [N/cm²]
400 / 0	1	1.74		0.18		44.44	55.44	1.18	0.91	
400 / 0	2	2.31		0.27		46.72	58.28	1.25	0.95	
400 / 0	2.5	2.50		0.33		48.33	60.29	1.29	0.98	
400 / 0	3	2.73		0.38		48.39	60.37	1.29	0.99	
400 / 0	3.8	3.00		0.45		50.52	63.02	1.35	1.03	
400 / 0	4	3.07		0.47		50.83	63.41	1.36	1.04	
400 / 0	4.5	3.23	2.5	0.51		50.95	63.64	1.36	1.04	
400 / 0	5	3.44	3.0	0.55	0.8	51.07	63.72	1.36	1.04	2.46
400 / 0	5.5	3.50	3.5 4.5	0.60	1.2	51.25	63.93	1.37	1.05	
400 / 0	5.9	3.55		0.64		51.51	64.26	1.37	1.05	
400 / 0	6.1	3.65		0.71		51.70	64.49	1.38	1.06	
400 / 0	6.4	3.74		0.79		52.13	65.04	1.39	1.06	
400 / 0	6.5	3.76		0.81		52.26	65.20	1.39	1.07	
400 / 0	6.6	3.77		0.83		52.27	65.21	1.39	1.07	
400 / 0	6.7	3.79		0.86		52.40	65.37	1.40	1.07	
400 / 0	7	3.83		0.94		52.87	65.95	1.41	1.08	

Track / Day	Factor	Comfort Index [-]	Comfort index limits	Y/Q [-]	Y/Q limit [-]	Q [kN]	Q_{max} [kN]	σ_z [N/cm²] 50 [cm]	60 [cm]	σ_{allow} [N/cm²]
400 / 0	7.5	3.95		1.06		53.38	66.60	1.42	1.09	
400 / 0	7.7	4.01	2.5	1.14		53.45	66.68	1.43	1.09	
400 / 0	7.8	4.02	3.0	1.17	0.8	53.53	66.77	1.43	1.09	2.46
400 / 0	7.9	4.04	3.5	1.22	1.2	53.58	66.84	1.43	1.09	
400 / 0	8	4.07	4.5	1.26		54.42	67.27	1.44	1.10	
400 / 0	9	4.42		1.49		53.65	66.63	1.42	1.09	

Table 45: **Magnitude increase to obtain limits of intervention that account for vehicle responses (own work)**

The normalized values form the able above can be visualized in Figure 103. Normalized values were used to compare the development of track parameter deterioration in order to determine the parameter that governs the limit determination.

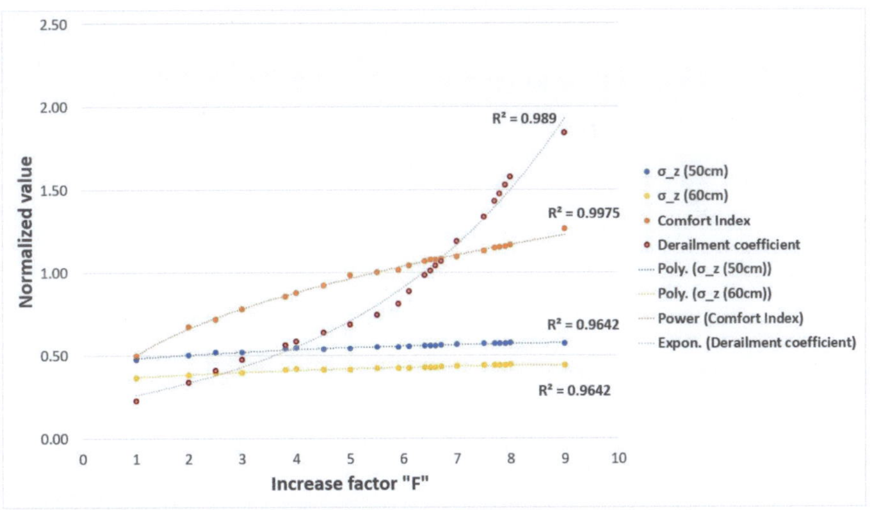

Figure 103: **Normalized values to determine the limit of intervention (own work)**

As seen in the figure, the normalized values extend beyond one since as discussed in section 5.3, the values were not normalized to the absolute maximum observed in the standard, but rather to a maximum value that does not correspond to a track posing extreme discomfort or high risk of derailment. In case of comfort, the maximum value was taken to be the value at which the vehicle experiences an uncomfortable ride. In case of safety, it was taken when the vehicle response equals a derailment coefficient of 0.8. In this sense, the absolute maximum values were considered to correspond to a track in which operation needs to be stopped.

Table 46 shows the normalized values corresponding to the limits (i.e. comfort, derailment coefficient and allowable stress) as established in Table 45 above.

Evaluation parameter	Description	Max values	Normalized values
Stress	σ_z	2.49	1.000
Comfort level	Middle comfort	2.50	0.714
	Half distance to uncomfortable	3.00	0.857
	Uncomfortable	3.50	1.000
	Very uncomfortable	4.50	1.286
Derailment coefficient	Max	0.80	1.000
	Absolute max	1.20	1.500

Table 46: **Normalized values using a maximum value not corresponding to the worst limit in the standards (own work)**

Appendix XIII: Flow Charts for Formal Specifications

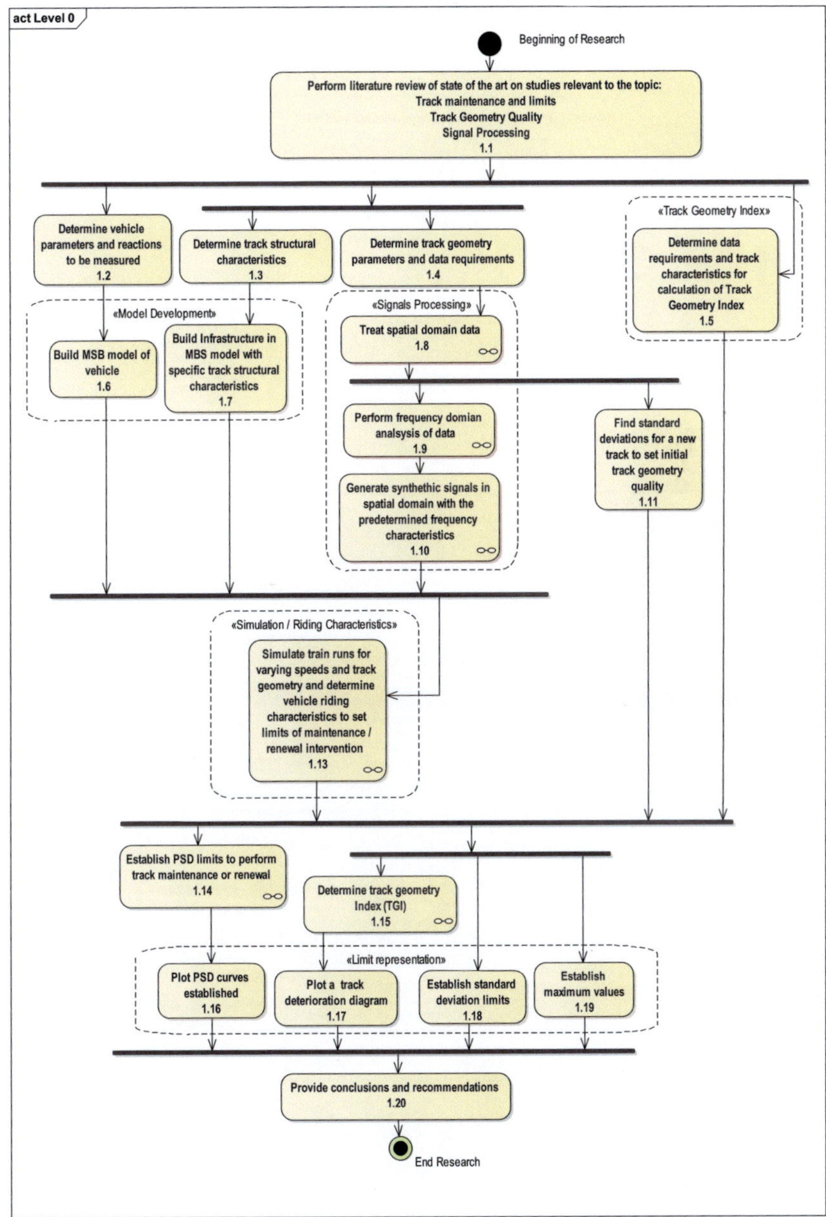

Figure 104: Overview of the research or higher-level processes to conduct the research (own work)

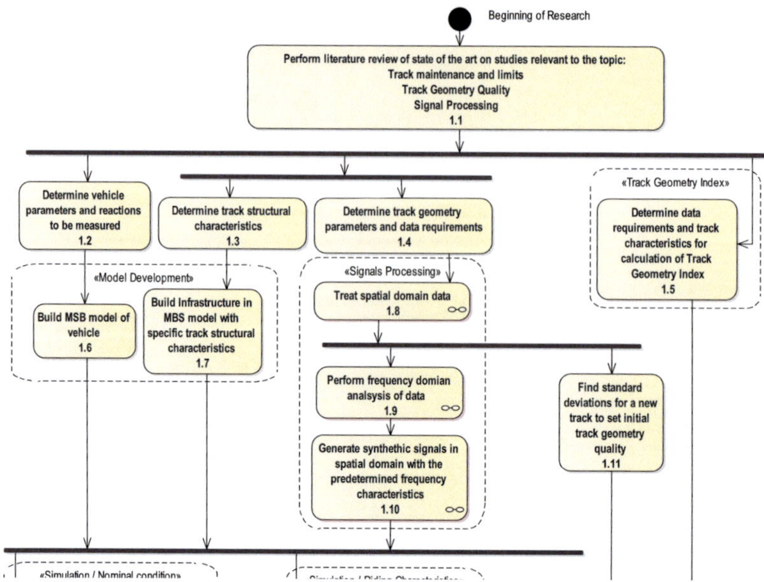

Figure 105: **Methodology applied for the development of the research (part 1) – (own work)**

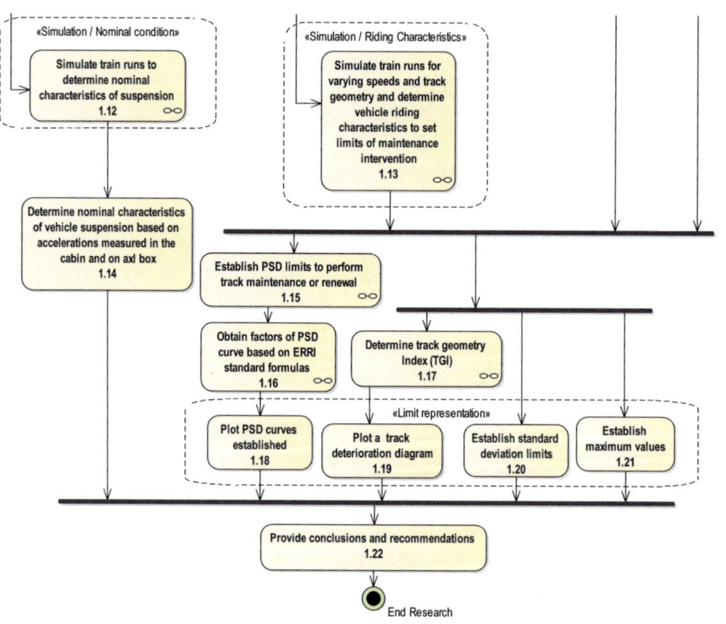

Figure 106: **Methodology applied for the development of the work (part 2) – (own work)**

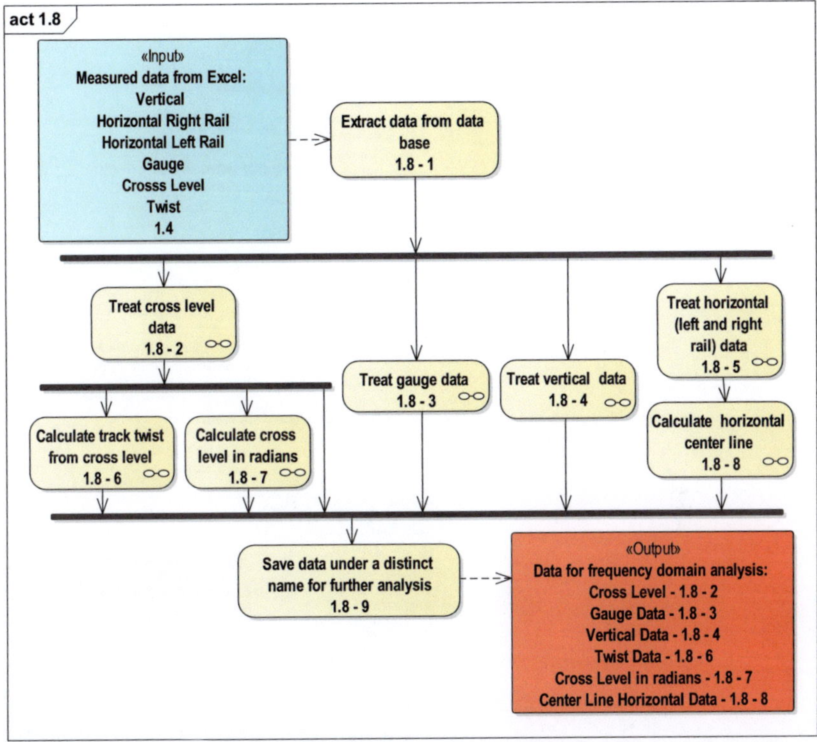

Figure 107:   Level 1.8 - Track data treatment (own work)

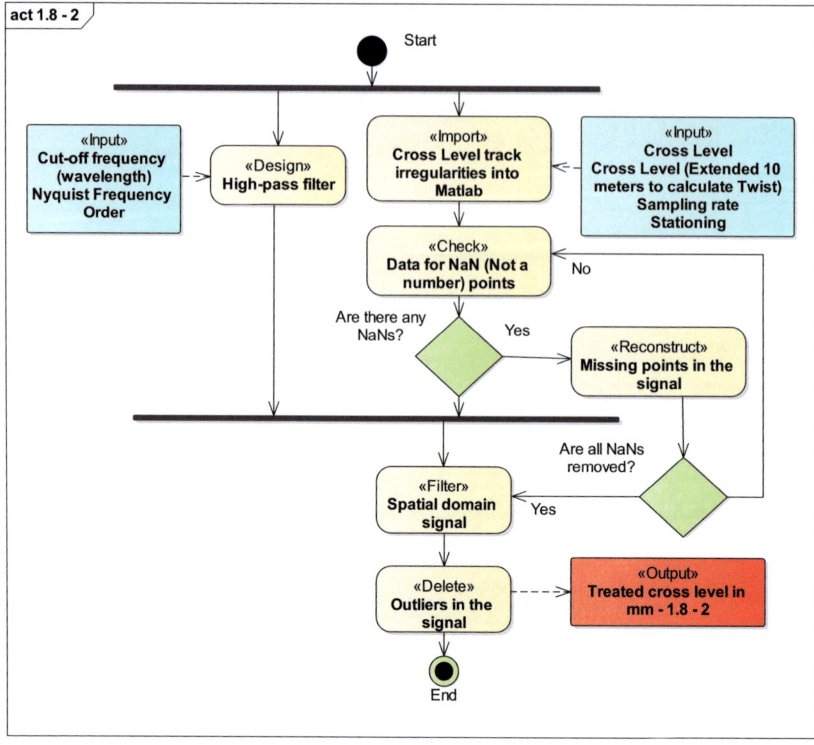

Figure 108:   **Cross Level data treatment (own work)**

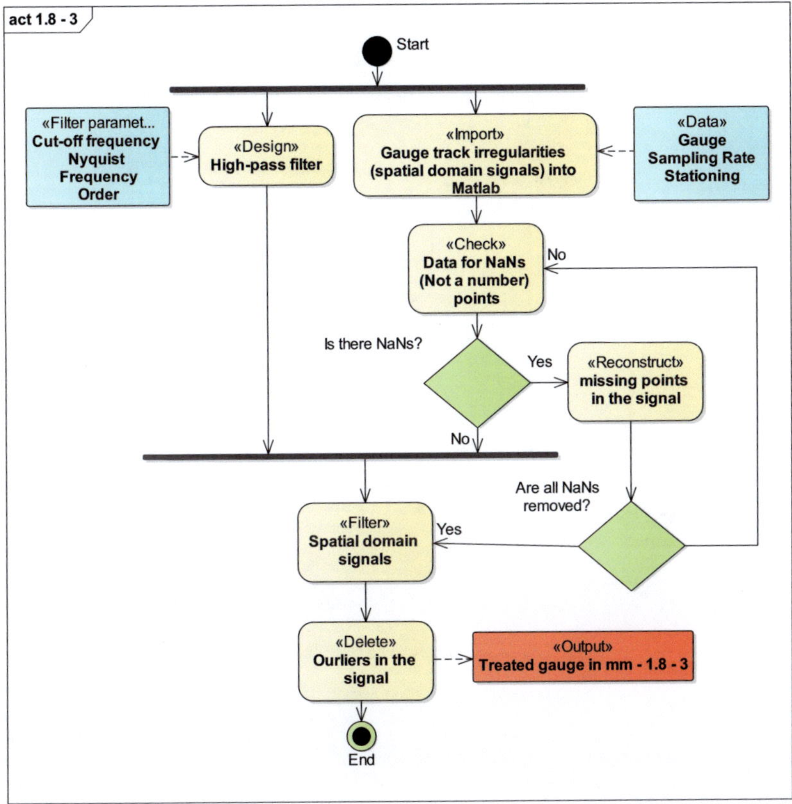

Figure 109: **Gauge data treatment (own work)**

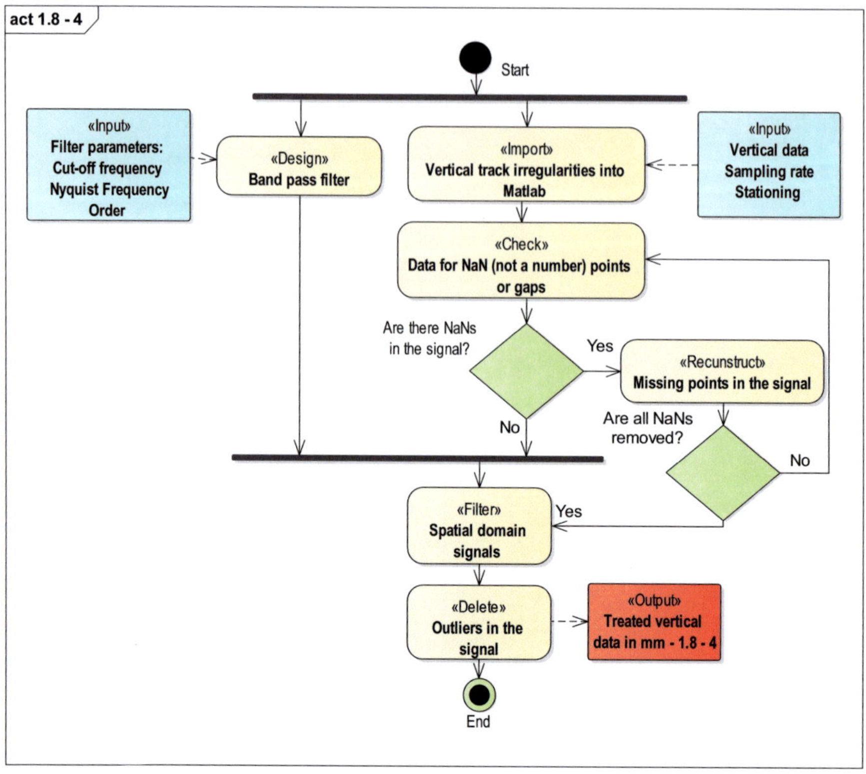

Figure 110:   Vertical data treatment (own work)

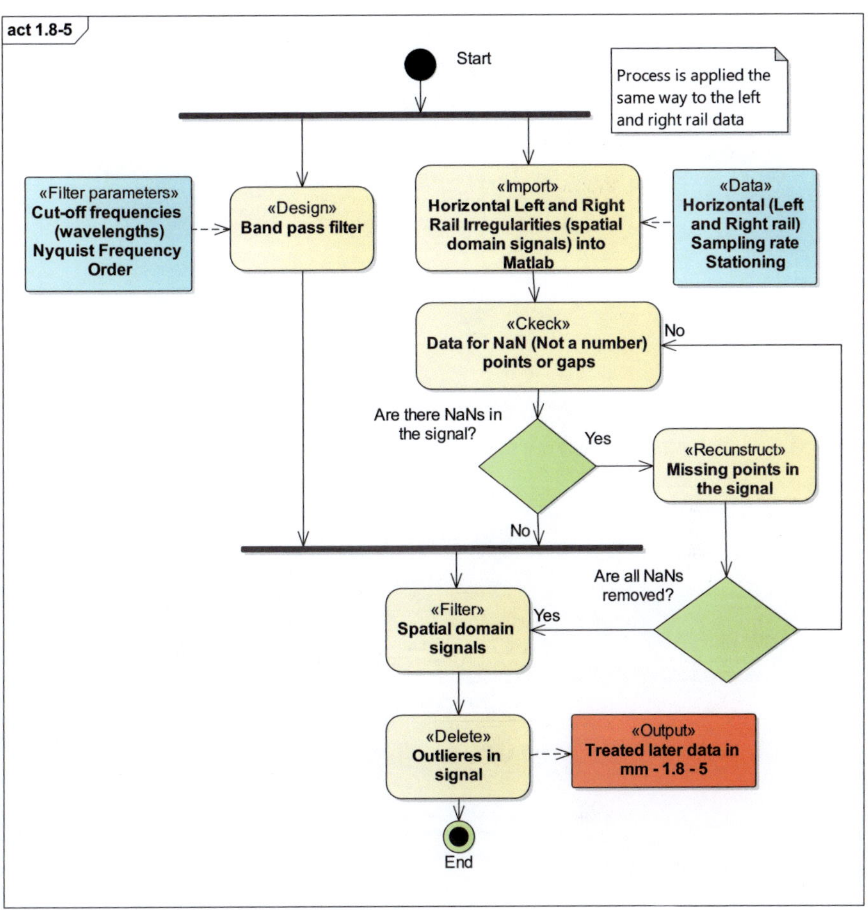

Figure 111:   **Horizontal data treatment (own work)**

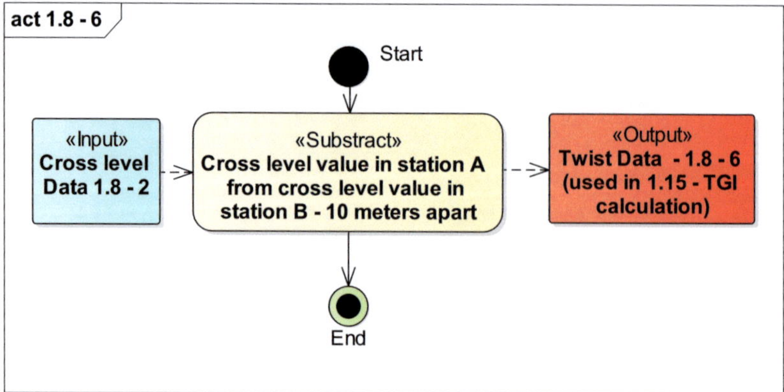

Figure 112:   **Calculation of track Twist (own work)**

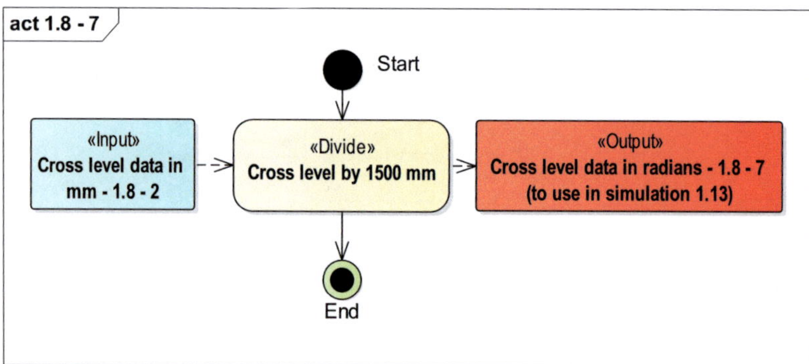

Figure 113:   **Calculation of Cross Level in radians (own work)**

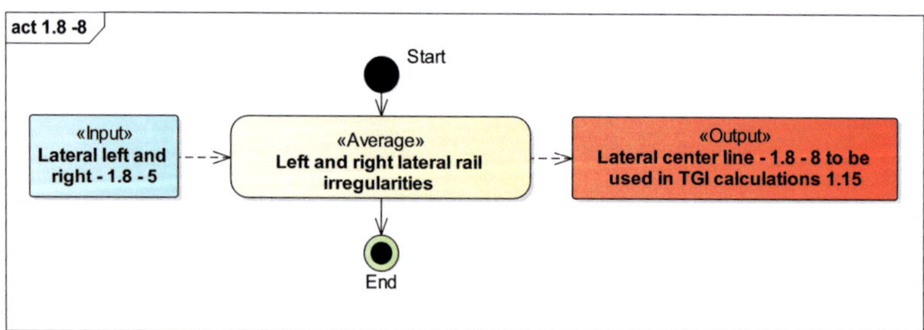

Figure 114:   **Calcualtion of Lateral Center line irregularities (own work)**

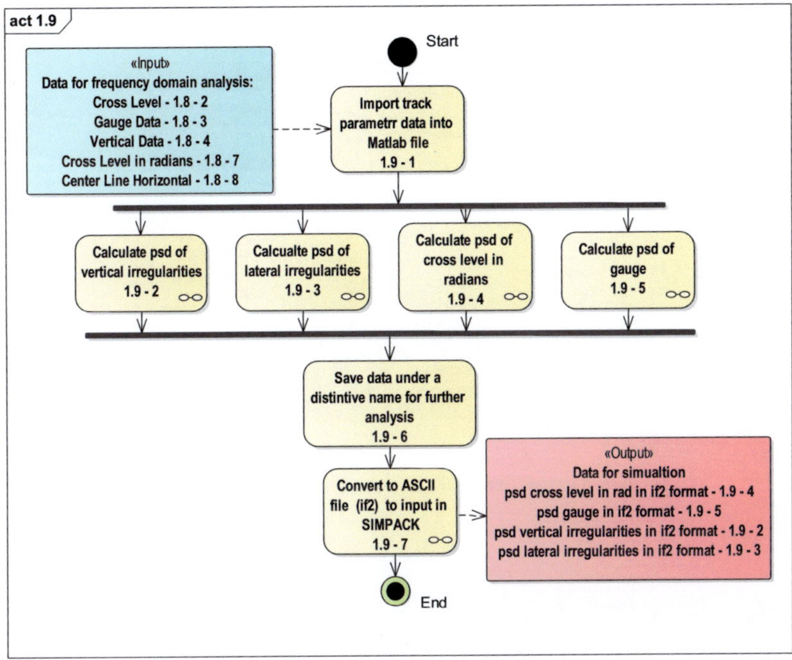

Figure 115:   Calculation of PSD of track irregularities (own work)

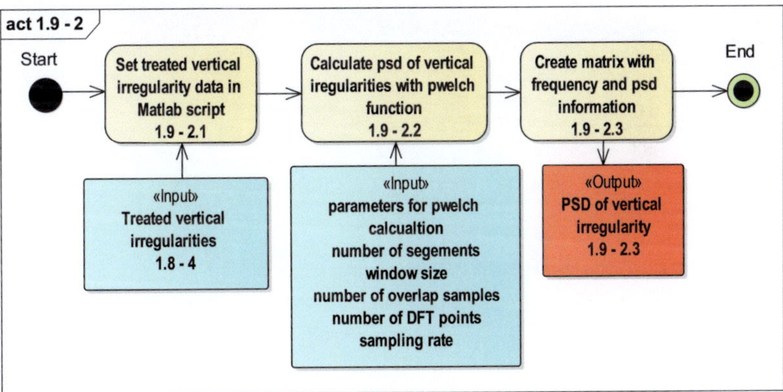

Figure 116:   Calculation of Vertical PSD (own work)

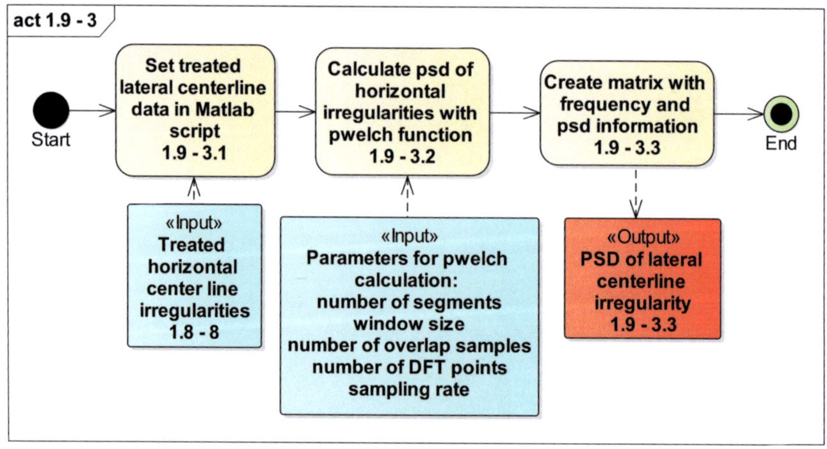

Figure 117: Calculation of Lateral Center line PSD (own work)

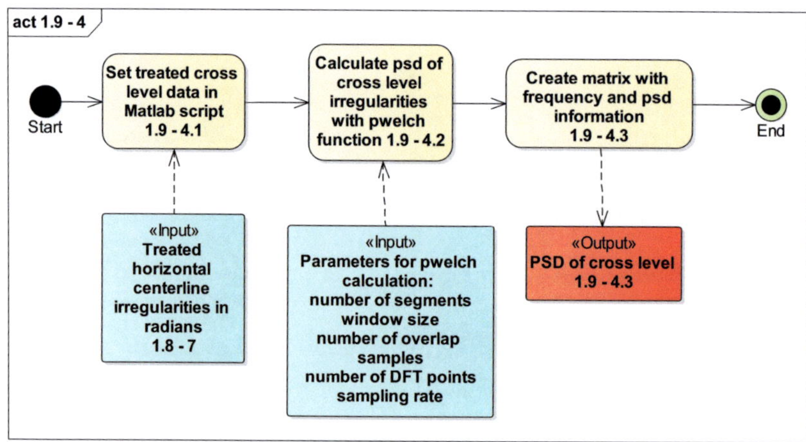

Figure 118: Calculation of Cross Level PSD (own work)

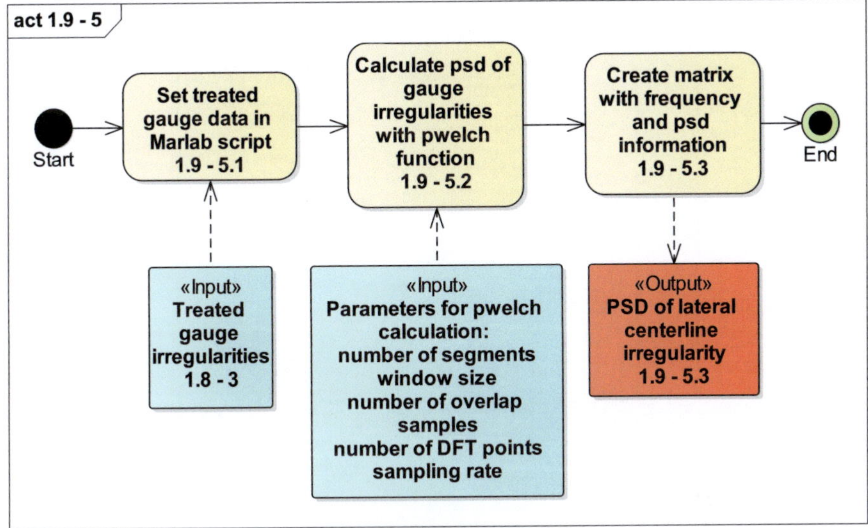

Figure 119: Calculation of Gauge PSD (own work)

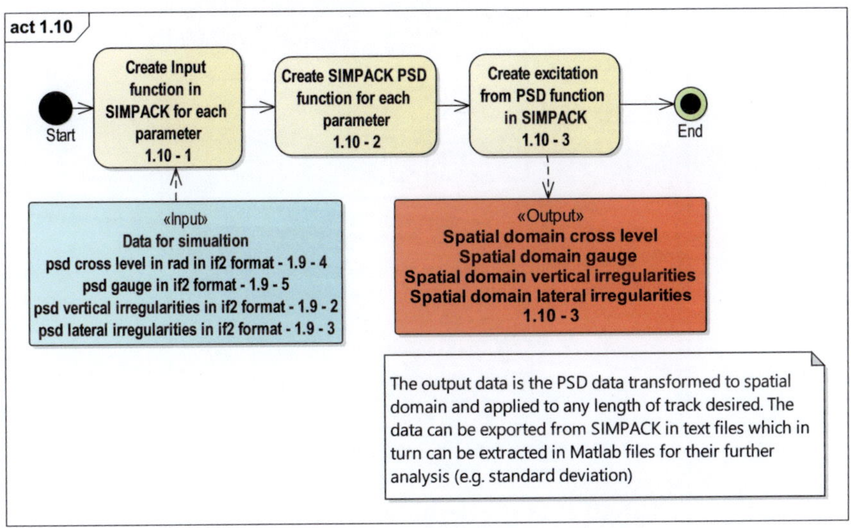

Figure 120: Determination of irregularities in PSD to use in SIMPACK® (own work)

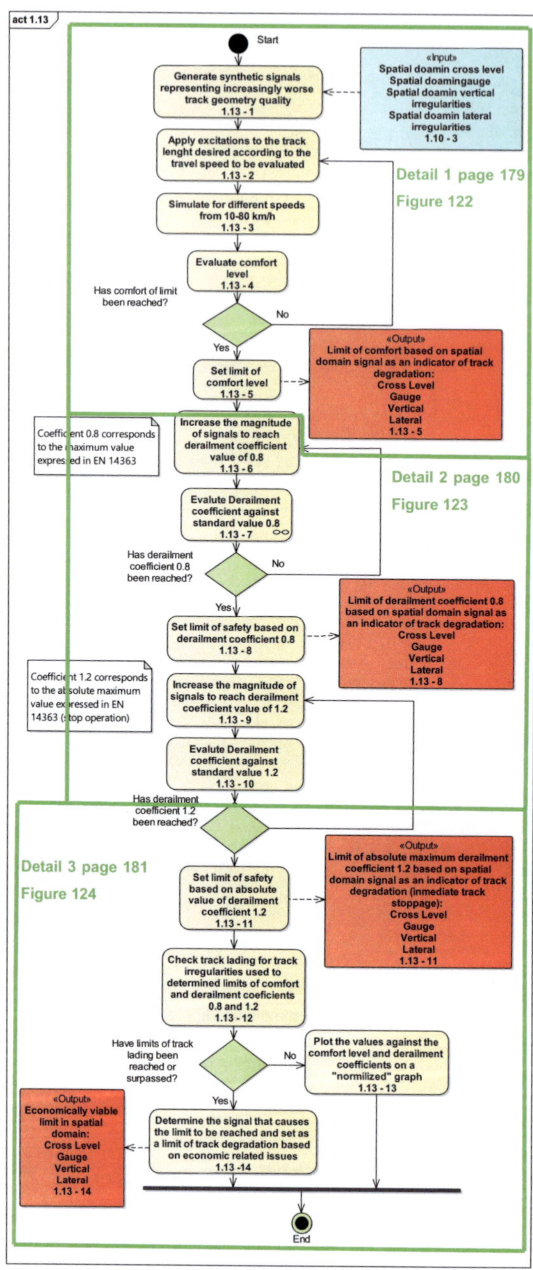

Figure 121: **Simulation of train runs for varying speeds and track geometry to determine vehicle riding characteristics**

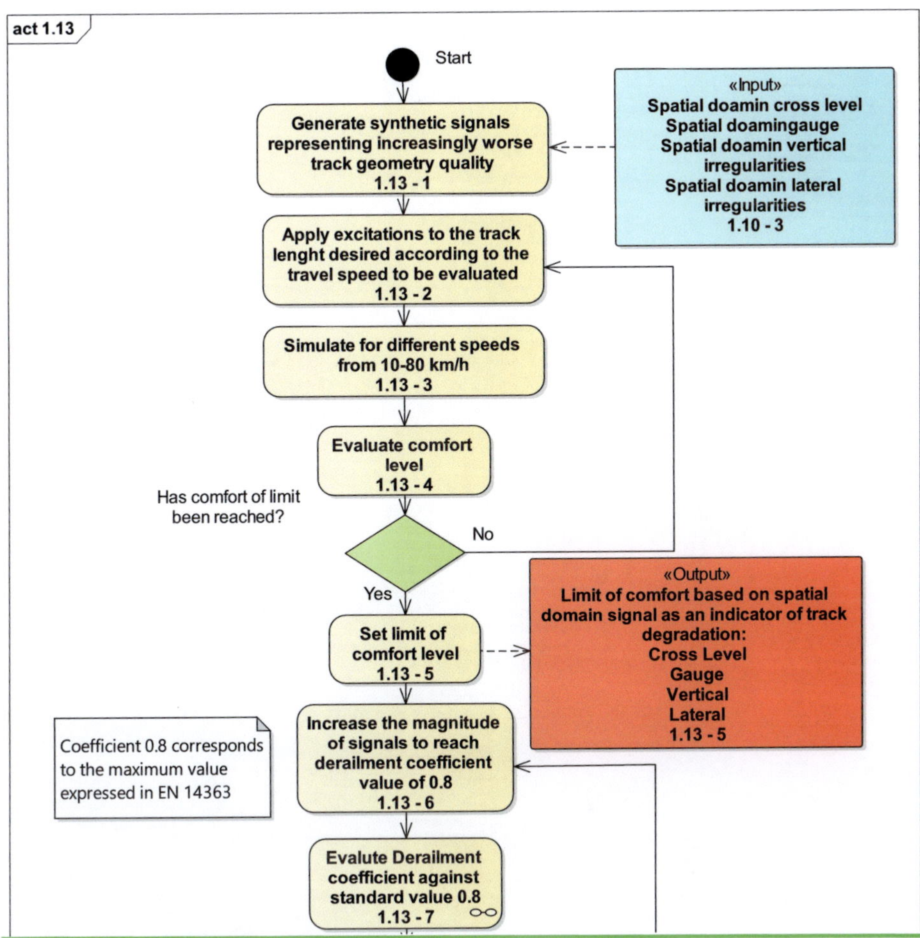

Continues on page 180 - Figure 123

Figure 122: Simulation of train runs for varying speeds and track geometry to determine vehicle riding characteristics – part 1 (own work)

Continued from page 179 - Figure 122

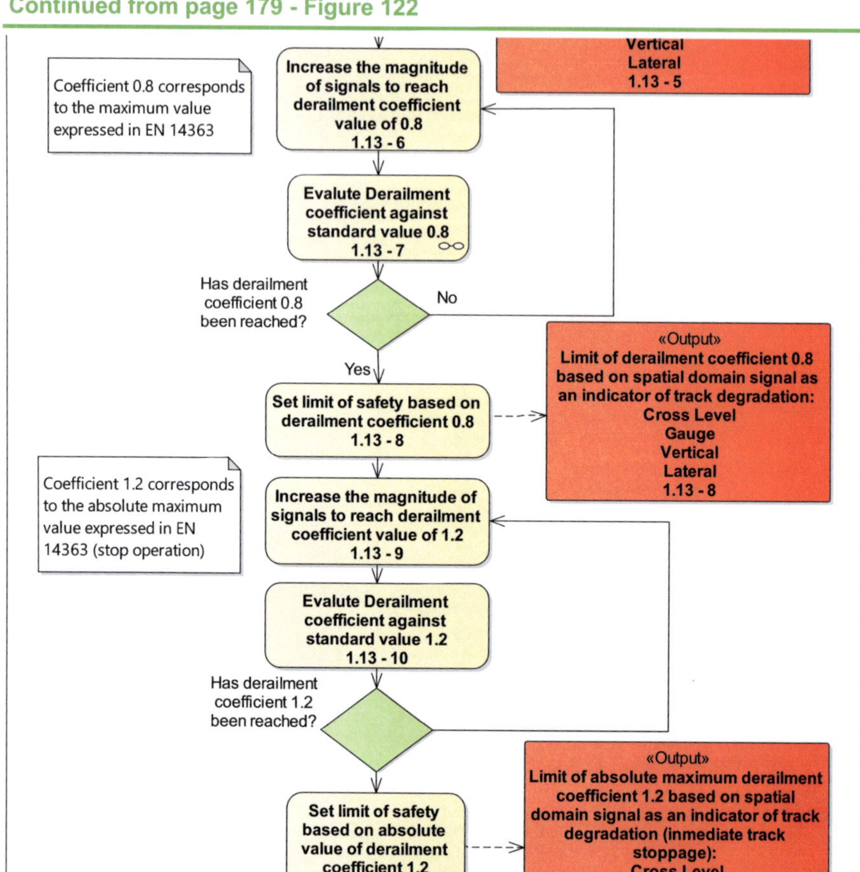

Continues on page 181 - Figure 124

Figure 123: **Simulation of train runs for varying speeds and track geometry to determine vehicle riding characteristics – part 2 (own work)**

Continued from page 180 - Figure 123

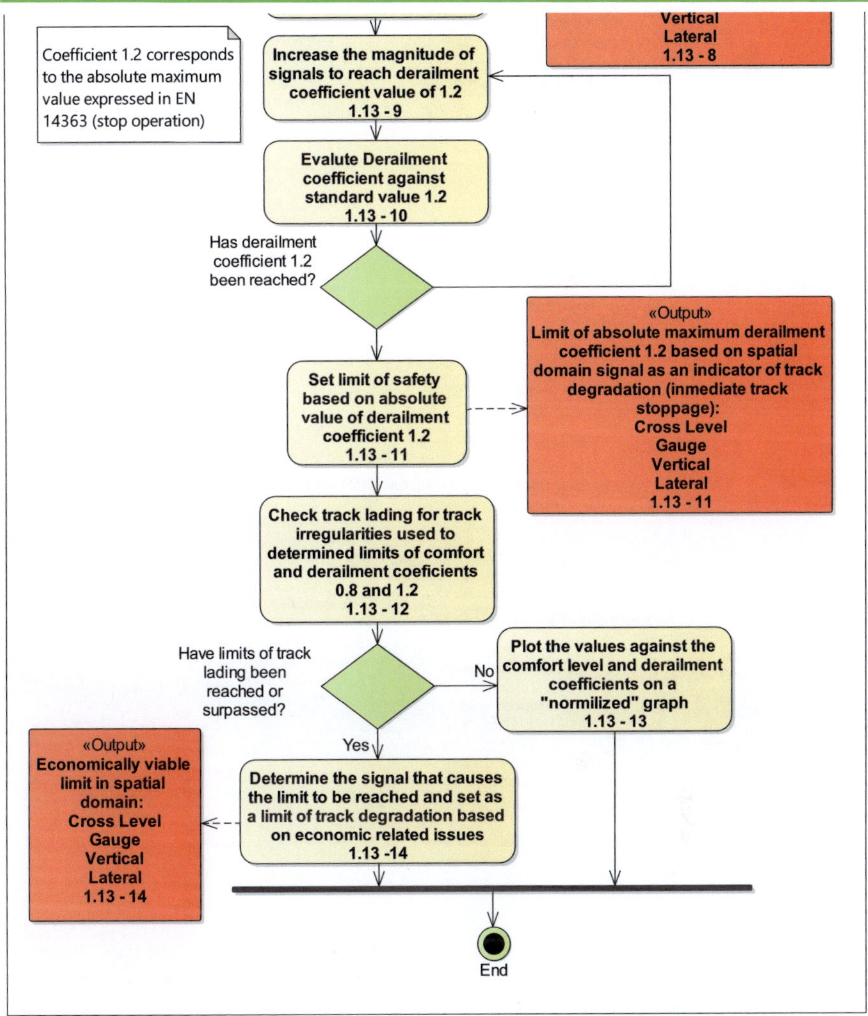

Figure 124: Simulation of train runs for varying speeds and track geometry to determine vehicle riding characteristics – part 3 (own work)

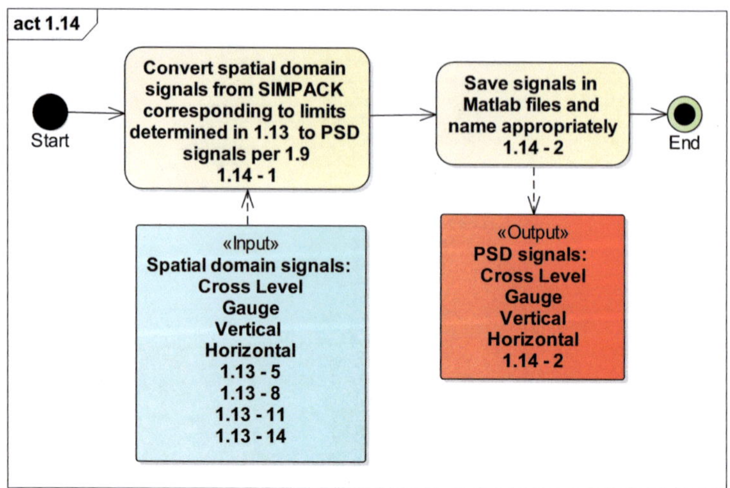

Figure 125: Conversion of spatial domain signals from SIMPACK© to PSD signals (own work)

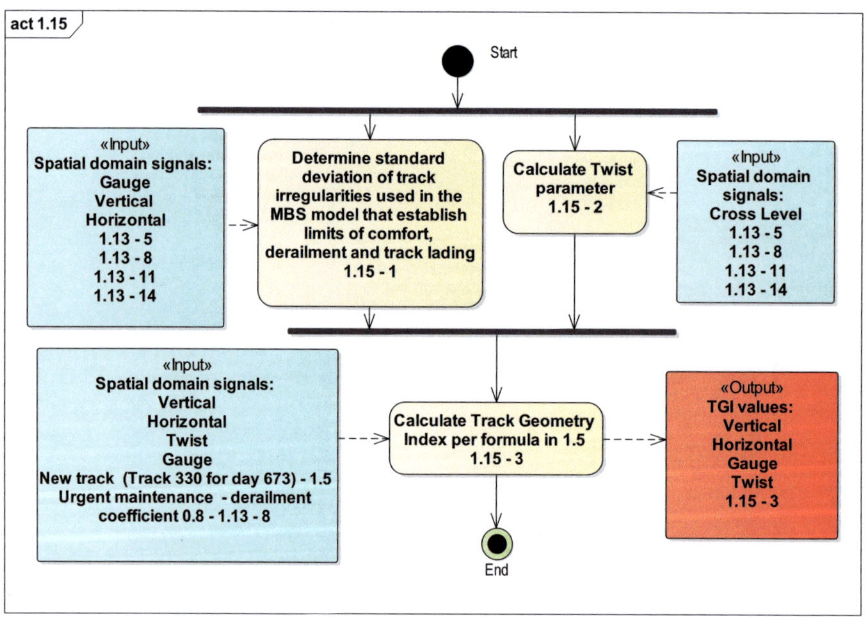

Figure 126: Calculation of track geometry index (own work)